ELECTRIC POWER SYSTEM BASICS FOR THE NONELECTRICAL PROFESSIONAL

SECOND EDITION

ELECTRIC POWER SYSTEM BASICS FOR THE NONELECTRICAL PROFESSIONAL

STEVEN W. BLUME

IEEE PRESS SERIES ON POWER ENGINEERING

IEEE PRESS

WILEY

Library of Congress Cataloging-in-Publication Data is available.

ISBN: 978-1-119-18019-7

Printed in the United States of America

SKY10061367_112923

CONTENTS

PREFACE

ABOUT THE BOOK

This book is intended to give nonelectrical professionals a fundamental understanding of large interconnected electrical power systems, better known as the "Power Grid" with regard to terminology, electrical concepts, design considerations, construction practices, industry standards, control room operations for both normal and emergency conditions, maintenance, consumption, telecommunications, and safety. Several practical examples, photographs, drawings, and illustrations are provided to help the reader gain a fundamental understanding of electric power systems. The goal of this book is to have the nonelectrical professional come away with an in-depth understanding of how power systems work from electrical generation to household wiring and consumption through connected appliances.

This book starts with terminology and basic electrical concepts used in the industry then progresses through generation, transmission, and distribution of electrical power. The reader is exposed to all the important aspects of an interconnected power system. Other topics discussed include energy management, conservation of electrical energy, consumption characteristics, and regulatory aspects to help readers understand modern electric power systems in order to effectively communicate with seasoned engineers, equipment manufacturers, field personnel, regulatory officials, lobbyists, politicians, lawyers, and others working in the electrical industry.

Please note that some sections within most chapters elaborate on certain concepts by providing additional details or background. These sections are marked "optional supplementary reading" and maybe skipped without losing value to the intent of this book.

This Second Edition provides updates to renewable energy (solar and wind primarily), equipment photo updates, updates in regulatory issues, new measures taken to improve system reliability, and more on smart technologies used in the power grid system.

ABOUT THE AUTHOR

The author, Steven W. Blume, is a registered professional engineer with a master's degree in electrical engineering and over 40 years' experience that covers all aspects of this book. Further, he has been teaching electric power system basics to nonelectrical professionals for over 25 years. His combined knowledge, experience, and ability to explain complex subjects in simple-to-understand terms present this book to those interested in gaining a fundamental understanding of electric power systems. Additional training materials are available to the reader such as online courses, instructor-led courses, private custom courses, and tour oriented courses. Visit www.aptc.edu. Blume is available at swblume@gmail.com

CHAPTER SUMMARIES

A brief overview of each chapter is presented below. Knowing where and when to expect specific topics and knowing how the information is organized in this book will help the reader comprehend the material easier. The language used in this book reflects actual industry terminology.

Chapter 1: System Overview, Terminology, and Basic Concepts

The book starts out with a brief yet informative discussion of the history that led to the power systems we know today. Then a system overview diagram is presented with brief discussions of the major divisions within an electric power system. Basic definitions and common terminology are then discussed such as voltage, current, power, and energy. Fundamental concepts such as direct and alternating current (i.e., dc and ac), frequency, single-phase and three-phase, types of loads, and power system efficiency are discussed to set the stage for more advanced learning.

How electrical generators produce electricity is also introduced in this chapter. The physical laws and concepts presented in this chapter serve as the foundation of all electric power systems.

Some very basic electrical formulas are presented in this chapter and at times elsewhere in the book. This is done intentionally to help explain terminology and concepts associated with electric power systems. The reader should not be intimidated nor concerned about the math, it is meant to describe and explain relationships.

Chapter 2: Generation

Basic concepts of the various electrical generation sources or power plants are presented in this chapter. These concepts include sub-systems that differentiate the plants regarding natural resources, spinning or non-spinning rotors, operational characteristics, environmental effects, and overall efficiencies.

The reader becomes more knowledgeable with the various aspects of electrical generation, the different prime movers used to rotate generator shafts and the basic building blocks that make up the various power plants. The prime movers discussed include steam, hydro, and wind turbines. Some of the non-rotating electric energy sources are also discussed in this chapter such as solar photovoltaic and biopower systems. The growth in renewable energy and its application to power grid operations are discussed.

The major equipment components or sub-systems associated with each power plant type are discussed such as boilers, cooling towers, boiler feed pumps, high- and low-pressure systems. The reader gains a basic understanding of power plant fundamentals as they read through this chapter.

Chapter 3: Transmission Lines

The reasons for using very high voltage power lines compared to low-voltage power lines are explained in this chapter. The fundamental components of transmission lines such as conductors, insulators, air gaps, and shielding are discussed. Direct current (dc) transmission and alternating current (ac) transmission lines are compared along with underground versus overhead. The reader will come away with a good understanding of transmission line design parameters and the benefits of using high-voltage transmission for efficient transport of electrical power.

Chapter 4: Substations

This chapter covers the equipment found in substations that transforms very high voltage electrical energy transported from generation facilities into a more useable form of electrical energy for distribution and consumption. The equipment themselves (i.e., transformers, regulators, circuit breakers, and disconnect switches) and their relationship to system protection, maintenance operations, and system control are discussed in this chapter. This chapter also includes discussions on new digital substation equipment being used to help modernize operations and reliability.

Chapter 5: Distribution

This chapter describes how primary distribution systems, both overhead and underground, are designed, operated, and used to serve residential, commercial, and industrial consumers. The distribution system between the substation and the consumer's demarcation point (i.e., service entrance equipment) is the focus. Overhead and underground line configurations, voltage classifications, and common equipment used in distribution systems are covered. The reader will learn how distribution systems are designed and built to provide reliable electrical power to the end users.

This chapter focuses on distribution systems in general; the modernization of distribution systems such as distribution automation, intelligent electronic devices, outage management, and customer information systems are discussed in Chapter 9.

Chapter 6: Consumption

The equipment located between the customer service entrance (i.e., demarcation point) and the wiring to the consumer's actual load devices are discussed in this chapter. The use of emergency generators and uninterruptible power supply (UPS) systems to enhance reliable power service are discussed along with their operating issues. Smart meters, service reliability indicators, common problems, and solutions associated with large power consumers are all covered in this chapter.

Chapter 7: System Protection

The difference between "system protection" and "personal protection" (i.e., safety) is explained first. Then this chapter is devoted to "system protection" and how electric power systems are protected against equipment failures, faults on power lines, lightning strikes, inadvertent operations, and other events that cause system disturbances. "Personal protection" is discussed later in chapter 10.

Reliable service is dependent upon properly designed and periodically tested protective relay systems. These systems, and their protective relays, are explained for transmission and distribution lines, substations, and generator units. The reader learns how the entire electric power system is designed to protect itself from power faults, lightning strikes, and to minimize the impact of major system disturbances.

Chapter 8: Interconnected Power Systems

This chapter starts out with a discussion of the four major power grids in North America and how these grids are territorially divided, operated, controlled, and regulated. The emphasis is to explain how the individual power companies are interconnected to improve the overall performance, reliability, stability, and security of the entire power grid. Other topics discussed include generation-load balance, resource planning, and operational limitations under normal and emergency conditions. Last, the concepts of rolling blackouts, brownouts, load shedding, and other service reliability issues and the methods used to minimize outages are discussed.

Chapter 9: System Control Centers and Telecommunications

System control centers are extremely important in the day-to-day operation of electric power systems. This chapter explains how system control center operators monitor and control equipment remotely. The advanced computer programs and telecommunications systems used to control power equipment remotely in substations, on power lines, and the actual consumer are discussed. These tools enable power system operators to economically dispatch power, meet system energy demands, control equipment during normal and during emergency conditions. The explanation and use of SCADA (Supervisory Control and Data Acquisition) and EMS (Energy Management Systems) are included in this chapter.

The functionality and benefits of the various types of communications systems used to connect system control centers with remote terminal units are discussed. These telecommunication systems include fiber optics, microwave, power line carrier, radio, and copper wireline circuits. The methods used to provide high speed protective relaying, customer service call centers, and digital data/voice/video communications services are all discussed in a fundamental manner.

The modernization of system control center tools such as synchrophasors and wide area monitoring systems to improve system security and reliability are discussed in this chapter. Distribution control and system modernization are also discussed in this chapter.

Chapter 10: Personal Protection (Safety)

The book concludes with a chapter devoted to electrical safety: personal protection and safe working procedures in and around high-voltage electric power systems.

Personal protective equipment such as rubber insulation products and the equipment necessary for effective grounding are described. Common safety procedures and proper safety methods are discussed, including "equipotential grounding." The understanding of "ground potential rise," "touch potential," and "step potential" adds a strong message to the reader as to the proper precautions needed when being around high-voltage power lines, substations, and even around the home.

This chapter includes a discussion around the very important issue of arc flash safety. The discussion includes governmental rules and regulations, proper safety procedures, responsibilities, and the special clothing needed to protect oneself from the hazard of arc flash should equipment unexpectedly explode or arc causing dangerous heat exposure when events like this are in close proximity to workers.

The last item discussed in this book is electrical safety around the home. Albeit high voltage is dangerous, normal residential voltage around the home is lethal too and safety around the home is also a very important topic to cover.

In summary, the purpose of this book is to give readers a basic overview of how electric power systems work, followed by a chapter on electrical safety around high-voltage equipment and the home.

ACKNOWLEDGMENTS

I WOULD personally like to thank several people who have contributed to the success of my career and the success of this book. To my wife Maureen who has been supporting me for well over 40 years. Thank you for your guidance, understanding, encouragement, and so much more. Thank you Michele Wynne; your enthusiasm, organizational skills, and creative ideas are greatly appreciated. Thank you Bill Ackerman; you are a great go-to person for technical answers, courseware development, and you always display professionalism and responsibility. Thank you John McDonald; your encouragement, vision, and recognition are greatly appreciated. I would also like to thank all of those who reviewed my final draft manuscript and provided professional suggestions that further enhanced this book for your benefit.

Steven W. Blume

CHAPTER 1

SYSTEM OVERVIEW, TERMINOLOGY, AND BASIC CONCEPTS

CHAPTER OBJECTIVES

After completing this chapter, the reader will be able to:

- ☑ *Discuss the history of electricity*
- ☑ *Explain the differences between voltage, current, power, and energy*
- ☑ *Describe how electricity is generated using nature's physical laws*
- ☑ *Describe the three types of load (electrical consumption) and their characteristics*
- ☑ *Discuss the three main components of a generator*

HISTORY OF ELECTRIC POWER

Benjamin Franklin is known for his discovery of electricity. Born in 1706, he began studying electricity in the early 1750s. His observations, including his kite experiment, verified the nature of electricity. He knew that lightning was very powerful and dangerous. The famous 1752 kite experiment had a pointed metal piece on the top and a metal key at the base end of the kite string. The string went through the key and attached to a Leyden jar. (A Leyden jar consists of two metal conductors separated by an insulator.) He held the string with a short section of dry silk as insulation from the lightning energy. He then flew the kite into a thunderstorm. He first noticed some loose strands of the hemp string stood erect, avoiding one another. (Hemp is a perennial American plant used in rope making by native Americans.) He proceeded to touch the key with his knuckle and received a small electrical shock.

Between 1750 and 1850, there were many great discoveries in the principles of electricity and magnetism by Volta, Coulomb, Gauss, Henry, Faraday, Tesla, and others. It was found that electrical current produces a magnetic field. And, it was found that a moving magnetic field near a wire produces electricity. This led to many

Electric Power System Basics for the Nonelectrical Professional, Second Edition. Steven W. Blume.
© 2017 by The Institute of Electrical and Electronics Engineers, Inc. Published 2017 by John Wiley & Sons, Inc.

inventions such as the battery (1800), generator (1831), motor (1831), telegraph (1837) and telephone (1876), plus many other intriguing inventions.

In 1879, Thomas Edison invented a more efficient light bulb similar to those in use today. In 1882, he placed into operation the historic Pearl Street steam-electric plant and the first direct current (dc) distribution system in New York City powering over 10,000 electric light bulbs. By the late 1880s, power demand for electric motors brought in 24-hour service and dramatically raised electricity demand for transportation and other industry needs. By the end of the 1880s, small centralized areas of electrical power distribution centers sprinkled the U.S. cities. Each distribution center was limited to a few blocks because of the transmission inefficiencies of using direct current. Voltage could not be increased or decreased using direct current systems and the need to transport power longer distances was in order.

To solve the problem of transporting electrical power long distances, George Westinghouse developed a device called the "transformer." The transformer allowed electrical energy to be transported long distances efficiently by raising the voltage to reduce losses. This made it possible to supply electric power to homes and businesses located far from the electric generating plants. The application of transformers required the distribution system to be of the alternating current (ac) type opposed to direct current (dc) type.

The development of the Niagara Falls hydroelectric power plant in 1896 initiated the practice of placing electric power generating plants far from consumption areas. The Niagara plant produced electricity to Buffalo, NY over 20 miles away. With Niagara, Westinghouse, using technology developed by Nicolas Tesla, who convincingly demonstrated the superiority of transporting power long distances with electricity using alternating current (ac) instead of direct current (dc). Niagara was the first large power system to supply multiple large consumers with only one power line across a long distance.

Since the early 1900s, alternating current power systems began appearing throughout the United States. These power systems became interconnected to form what we know today as four major power grids in the United States and Canada.

It is interesting to note, however that direct current systems are coming back. For example, rooftop solar, dc transmission lines, and other dc generation and load devices are growing at a significant rate.

The remainder of this chapter discusses the fundamental terms and concepts used in today's electric power systems based on this impressive history.

SYSTEM OVERVIEW

Electric power systems are real-time energy delivery systems. Real-time meaning power is generated, transported, and supplied the moment you turn on the light switch. Electric power systems are not storage systems like water systems and gas systems. Instead, generators produce the energy as the demand calls for it!

Figure 1-1 shows the basic building blocks of an electric power system. Starting with *generation*, where electrical energy is produced in the power plant and then transformed in the power station to high-voltage electrical energy that is more suitable

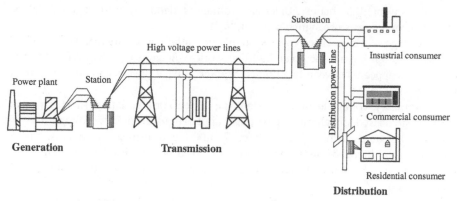

Figure 1-1   System overview.

for efficient long-distance transportation. The power plants transform other sources of energy as well in the process of producing electrical energy. For example, heat, mechanical, hydraulic, chemical, solar, wind, geothermal, nuclear, and other energy sources are used in the production of electrical energy. High-voltage (HV) power lines in the *transmission* portion of the electric power system efficiently transport electrical energy long distances to the consumption locations. Finally, the remote substations are responsible for transforming this HV electrical energy for delivery on lower high-voltage power lines called "Feeders" that are more suitable for the *distribution* of electrical energy. This electrical energy is again transformed to even lower voltage services for residential, commercial, and industrial consumption.

A full-scale actual interconnected electric power system is much more complex than that shown; however, the basic principles, concepts, theories, and terminologies are all the same. We will start with the basics and add complexity as we progress through the material.

TERMINOLOGY

Let us start with building a good understanding of the basic terms and concepts most often used by industry professionals and experts to describe and discuss electrical issues in small-to-large power systems. Please take the time necessary to grasp these basic terms and concepts. We will use them throughout this book to build a complete working knowledge of electrical power systems.

Voltage

The first term or concept to understand is *voltage*. Voltage is the *potential energy* source in an electrical circuit to make things happen. It is sometimes called *electromotive force* or EMF. The unit of Voltage is the **Volt**. The Volt was named in honor of Allessandro Giuseppe Antonio Anastasio Volta (1745–1827), the Italian physicist who also invented the battery. Electrical voltage is identified by the symbol "e" or "E" (some references use symbols "v" or "V").

Voltage is the electric power system's potential energy source. Voltage does nothing by itself but has the potential to do work. Voltage is a push or a force. Voltage always appears between two points. Voltage is what pushes and pulls electrons through wires.

Normally, voltage is either constant (i.e., direct) or alternating. Electric power systems are based on alternating voltage applications from low voltage 120 V residential systems to ultra-high voltage 765,000 V transmission systems. There are lower- and higher-voltage applications involved in electric power systems, but this is the range commonly used to cover generation through distribution and consumption.

In water systems, voltage corresponds to the pressure that pushes water through a pipe. Similar to voltage in wires, pressure is present in water pipes even though no water is flowing!

Current

Current is the flow of electrons in a *conductor* (wire). Electrons are pushed and pulled by voltage through an *electrical circuit* or closed loop path. The electrons flowing in a conductor always return to their voltage source. The unit of Current is *ampere* (also called amps), named after Andre-Marie Ampere, a French physicist. (One ampere is equal to 628×10^{16} electrons flowing in the conductor per second.) The number of electrons never decreases in a loop or circuit. The flow of electrons in a conductor produces heat from the conductor's *resistance* (i.e., friction).

Voltage always tries to push or pull current. Therefore, when a complete circuit or closed loop path is provided, voltage will cause current to flow. The resistance in the circuit will reduce the amount of current flow and will cause heat to occur. The *potential energy* of the voltage source is hereby converted into *kinetic energy* as the electrons flow. The kinetic energy is then utilized by the *load* (i.e., consumption device(s)), where it is converted into useful work.

Current flow in a conductor is similar to ping-pong balls lined up in a tube. Referring to Figure 1-2, a pressure on one end of the tube (i.e., voltage) pushes the balls through the tube. The pressure source (i.e., battery) collects the balls exiting the tube and re-enters them into the tube in a circulating manner (closed loop path).

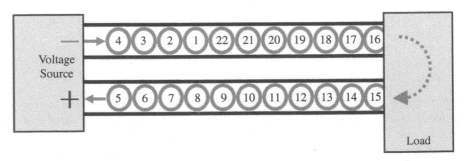

Figure 1-2   Current flow.

The number of balls traveling through the tube per second is analogous to current. This movement of electrons in a specified direction is called *current*. Electrical current is identified by the symbol "*i*" or "*I*."

Hole Flow vs. Electron Flow

Electron flow is when electrons go from one atom to the next while moving toward the positive side of the voltage source. As an electron leaves one atom and goes to the next, it leaves a hole or vacancy behind. The holes left behind can be thought of as a current of vacancies moving toward the negative side of the voltage source. Therefore, as electrons flow in a circuit one direction, holes are created in the same circuit that flow in the opposite direction. Current is defined as either electron flow or hole flow. *The standard convention used in electrical circuits is hole flow!* (One reason for this is that the concept of positive (+) and negative (−) terminals on a battery or voltage source was established long before the electron was discovered. The early experiments simply defined current flow as being from positive to negative, without really knowing what was actually moving!)

One important phenomenon about current flowing in a wire that we will discuss in more detail later is the fact that "*a current flowing in a conductor produces a magnetic field!*" See Figure 1-3. This is a physical law, similar to gravity being a physical law. For now, just keep in mind that when electrons are pushed or pulled through a

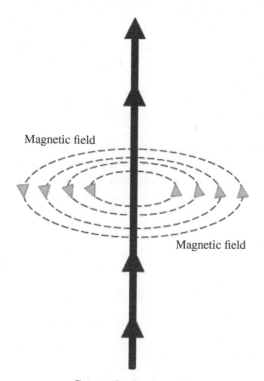

Magnetic field

Magnetic field

Current flowing in a wire

Figure 1-3   Current and magnetic field.

wire by voltage, a magnetic field is produced automatically around the wire. Note: Figure 1-3 is a diagram that corresponds to the direction of conventional or hole flow current according to the "right hand rule."

Power

The unit of *power* is the *Watt*, named after James Watt (1736–1819), also the inventor of the steam engine. Voltage by itself does not do any real work. Current by itself does not do any real work. However, voltage and current together can produce real work. The product of voltage and current is power. Power is used to produce real work.

For example, electrical power can be used to create heat, spin motors, light lamps, etc. The fact that power is part voltage and part current is that power equals zero if either voltage or current is zero. Voltage might appear at a wall outlet in your home and a toaster plugged into the outlet, but until someone turns on the toaster no current flows, and hence no power occurs until the switch is turned on and current is flowing through the high-resistive wires creating heat.

Energy

Electrical *energy* is the product of electrical power and time. The amount of time a load is on (i.e., current is flowing) times the amount of power used by the load (i.e., Watts) is energy. The measurement for electrical energy is *watt-hours*. The more common units of energy in electric power systems are kilowatt-hours (kWh, meaning 1000 watt-hours) for residential applications and megawatt-hours (MWh, meaning 1,000,000 watt-hours) for the large industrial applications or the power companies themselves.

DC Voltage and Current

Direct current (dc) is the flow of electrons in a circuit that is always in the same direction. Direct current (i.e., one direction current) occurs when the voltage is kept constant, as shown in Figure 1-4. A battery, for example, produces dc current when connected to a circuit. The electrons leave the negative terminal of the battery and move through the circuit toward the positive terminal of the battery. The holes, however, flow in the opposite direction.

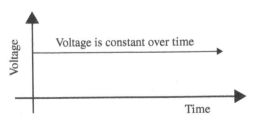

Figure 1-4   Direct (i.e., dc voltage).

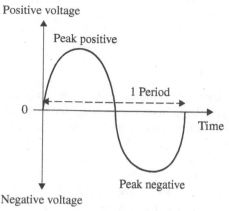

Figure 1-5   Alternating (i.e., ac voltage).

AC Voltage and Current

When the terminals of the potential energy source (i.e., voltage) alternate positive and negative, the current flowing in the electrical circuit likewise alternates positive and negative (or clockwise and counterclockwise in the closed loop path). Thus alternating current (ac) occurs when the voltage source alternates.

Figure 1-5 shows the voltage increasing from zero to a positive peak value then decreases through zero to a negative peak value and back through zero again completing one cycle or in mathematical terms; this describes a *sine wave*. The sine wave can repeat many times in a second, minute, hour, or day. The length of time it takes to complete one cycle in a second is called the *period* of the cycle.

Comparing AC and DC Voltage and Current

Electrical load such as light bulbs, toasters, and hot water heaters can be served by either ac or dc voltage and current. However, dc voltage sources continuously supply heat in the load while ac voltage sources cause heat to increase and decrease during the positive part of the cycle, then increase and decrease again in the negative part of the cycle. In ac circuits, there are actually moments of time when the voltage and current are zero and no additional heating occurs.

It is important to note that there is an equivalent ac voltage and current that will produce the same heating effect in electrical load as if the source were dc voltage and current. The equivalent voltages and currents are referred to as the *root mean squared* values, or *rms* values. The reason this concept is important to understand is that all electric power system equipment (including HV power lines) are rated in rms voltages and currents.

For example, the 120 Vac wall outlet is actually the rms value. Theoretically, one could plug a 120 Vac toaster into a 120 Vdc battery source and cook the toast in the same amount of time. The ac rms value has the same heating effect as its equivalent dc value.

(Optional Supplementary Reading)
Appendix A explains how rms is derived.

Frequency

Frequency is the term used to describe the number of sine wave cycles in a second. The number of cycles per second is also called *Hertz*. Hertz was named after Heinrich Hertz (1857–1894) a German physicist. Note: direct current (dc) has no frequency, therefore, frequency is a term used only for ac circuits.

For electric power systems in the United States, the standard frequency is 60 cycles/second or 60 Hz. The European countries have adopted 50 Hz as the standard frequency. Countries outside the United States and Europe use 50 and/or 60 Hz. (Note at one time the United States had 25, 50, and 60 Hz systems. These were later standardized to 60 Hz.)

Phase Angle

In ac power systems, the voltage and current have the same frequency but have different amplitudes and phase angles. The *phase angle* between voltage and current is shown in the Figure 1-6. Note in this figure the current wave crosses the horizontal axis after the voltage wave and therefore is said current lags the voltage. Load devices that make current lag voltage are considered *inductive* (more on this later).

Note too that the amplitude of current at the same time voltage peaked is less than the peak current. This difference in current amplitude has great significance when it comes to minimizing power losses and maximizing overall power system *efficiency*. In other words, *reducing the phase angle reduces the amount of current needed to get the same amount of work done by the loads.* For example, adding capacitors (leading devices, which behave opposite to inductors and discussed in more detail later) to motors reduces the total current required from the generation source. Reducing the total current reduces system losses and improves the overall efficiency of the power system.

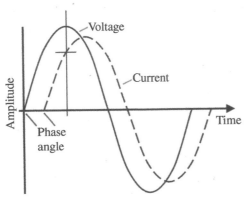

Figure 1-6   Phase angle between voltage and current.

AC VOLTAGE GENERATION

There are basically two physical laws that describe how electric power systems work. (Gravity is an example of a physical law.) One law has to do with generating a voltage from a changing magnetic field and the other has to do with a current flowing through a wire creating a magnetic field. Both physical laws are used throughout the entire electric power system from generation through transmission, distribution, and consumption. The combination of these two laws makes our electric power systems work. Understanding these two physical laws will enable the reader to fully understand and appreciate the fundamental concepts behind electric power systems operation.

Physical Law #1

AC voltage is generated in electric power systems by a very fundamental physical law called *Faraday's Law*. Faraday's Law represents the phenomena behind how electric motors turn and how electric generators produce electricity. Faraday's Law is the foundation for electric power systems.

Faraday's Law states, "A voltage is produced on any conductor in a changing magnetic field." It may be difficult to grasp the full meaning of that statement at first. It is however easier to understand the meaning and significance of this statement through graphs, pictures, and animations.

In essence, this statement is saying that if one takes a coil of wire and puts it next to a moving or rotating magnet a measurable voltage will be produced in that coil. Generators, for example, use a spinning magnet (i.e., rotor) next to a coil of wire to produce voltage. This voltage is then distributed throughout the electric power system.

We will now study how a generator works. Keep in mind that virtually all generators in service today have coils of wire mounted on stationary housings, called *stators*, where voltage is produced due to the changing *magnetic field* provided on the spinning *rotor*. The rotor is sometimes called the *field* because it is responsible for the magnetic field portion of the generator. The rotor's strong magnetic field passes the stator windings (coils), thus producing or generating an alternating voltage (ac) in the stator wires that is based on Faraday's Law. This principle will be shown and described in the following sections.

The amplitude of the generator's output voltage can be changed by changing the strength of rotor's magnetic field. Thus, the generator's output voltage can be lowered by reducing the rotor's magnetic field's strength. The means, by which the magnetic field in the rotor is actually changed will be discussed later in this book when Physical Law #2 is discussed.

Single-Phase AC Voltage Generation

Placing a coil of wire (i.e., conductor) in the presence of a moving magnetic field produces a voltage as discovered by Faraday. This principle is graphically presented in Figure 1-7. While reviewing the drawing, note that changing the rotor's speed

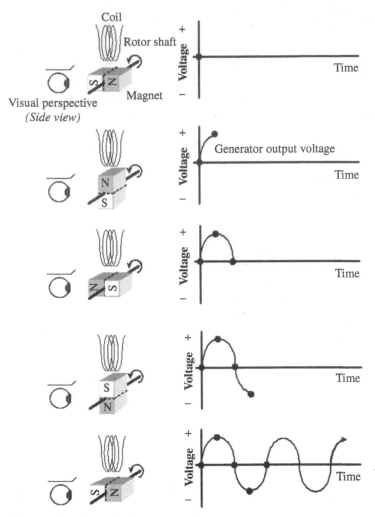

Figure 1-7   Magnetic sine wave.

changes the frequency of the sine wave. Also recognize the fact that increasing the number of turns (loops) of conductor or wire in the coil increases the resulting output voltage.

Three-Phase AC Voltage Generation

When *three* coils of wire are placed in the presence of a changing magnetic field, three independent voltages are produced. When the coils are spaced 120 degrees apart in a 360-degree circle, *three-phase* ac voltage is produced. As shown in Figure 1-8, three-phase generation can be viewed as three separate single-phase generators, each of which are displaced 120 degrees, and all of which share the same rotor's magnetic field.

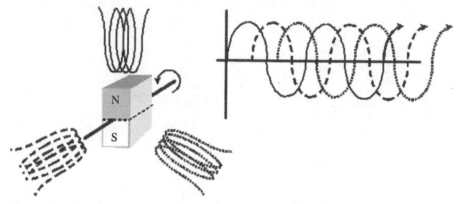

Figure 1-8   Three-phase voltage production.

The Three-Phase AC Generator

Large and small generators that are connected to the power grid system have three basic components; stator, rotor, and exciter. This section discusses these three basic components.

The Stator

A three-phase ac generator has three single-phase windings. These three windings are mounted on the stationary part of the generator, called the *stator*. The windings are physically spaced so that the changing magnetic field presented on each winding is 120 degrees out of phase with the other windings. A simplified drawing of a three-phase generator is shown in Figure 1-9.

The Rotor

The *rotor* is the center component that when turned moves the magnetic field. A rotor could have a *permanent magnet* or an *electromagnet* and still function as a generator. Large power plant generators use electromagnets so that the magnetic field strength

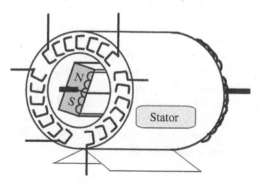

Figure 1-9   Three-phase generator–stator.

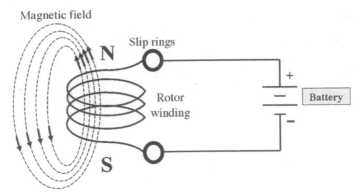

Figure 1-10   Electromagnet and slip rings.

can be varied. Varying the magnetic field of the rotor enables generation control systems to adjust the output voltage according to load demand and system losses. A drawing of an electromagnet is shown in Figure 1-10.

The operation of electromagnets is described by Physical Law #2.

Ampere's Law and Lenz's Law (Physical Law #2)   The second basic physical law that explains how electric power systems work is the fact that current flowing in a wire produces a magnetic field. Ampere's and Lenz's law state that "*a current flowing in a wire produces a magnetic field around the wire.*" These laws together describe the relationship between the production of magnetic fields and electrical current flowing in a wire. In essence, when current flows through a wire, a magnetic field surrounds the wire.

Electromagnets   Applying a voltage (e.g., battery) to a coil of wire produces a magnetic field. The coil's magnetic field will have a north and a south pole as shown in Figure 1-10. Increasing the voltage or the number of turns in the winding increases the magnetic field. Conversely, decreasing the voltage or number of turns in the winding decreases the magnetic field. *Slip rings* are electrical contacts that are used to connect the stationary battery to the rotating rotor as shown in Figure 1-10.

Rotor Poles   Increasing the number of magnetic poles on the rotor enables rotor speeds to be slower and still maintain the same electrical output frequency. Generators that require slower rotor speeds to operate properly use multiple pole rotors. For example, hydropower plants use generators with multiple pole rotors because the prime mover (i.e., water) is very dense, moves relatively slow compared to high-pressure steam turbines and harder to control than light-weight steam.

The relationship between the number of poles on the rotor and the speed of the shaft is determined using the following mathematical formula:

$$\text{Revolutions per minute} = \frac{7200}{\text{Number of poles}}$$

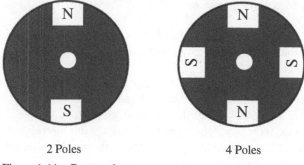

2 Poles   4 Poles

Figure 1-11   Rotor poles.

Figure 1-11 shows the concept of multiple poles in a generator rotor. Since these poles are derived from electromagnets, having multiple windings on a rotor provide the multiple poles.

Example 1. A two-pole rotor would turn 3600 rpm for 60 Hz.

Example 2. Some of the generators at Hoover Dam near Las Vegas, Nevada, use 40-pole rotor. Therefore, the rotor speed is 180 rpm or 3 revolutions per second, yet the electrical frequency is still 60 cycles/second (or 60 Hz). One can actually see the shaft turning at this relatively slow rotational speed.

The Exciter

The voltage source to the rotor that eventually creates the rotor's magnetic field is called the *exciter* and the coil on the rotor is called the *field*. Figure 1-12 shows

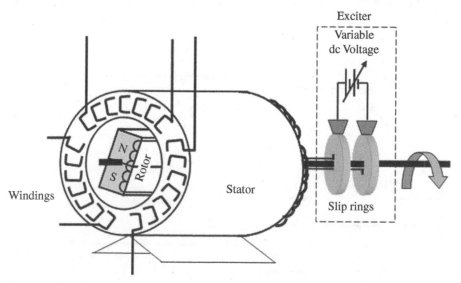

Figure 1-12   Three-phase voltage generator components.

the three main components of a three-phase ac generator; the stator, rotor, and exciter.

Figure 1-12 shows the *slip rings* used to complete the circuit between the stationary exciter voltage source and the rotating coil on the rotor, where the electromagnet produces the north and south poles.

Note: adding load to a generator's stator windings reduces rotor speed because of the repelling forces between the stator's magnetic field and the rotor's magnetic field, since both windings have electrical current flowing through them. Conversely, removing load from a generator increases rotor speed. Therefore, the *mechanical energy* of the prime mover that is responsible for spinning the rotor must be adjusted to maintain rotor speed or frequency under varying load conditions.

AC CONNECTIONS

There are two ways to connect the three windings that have a total of six leads (the ends of the winding wires) symmetrically. The two symmetrical connection configurations of a three-phase generator (or motor) are called *delta* and *wye*. Figure 1-13 shows these two connection types. Generators usually have their stator winding connected internally in either a delta or wye configuration.

The generator *nameplate* specifies which winding configuration is used on the stator.

Delta

Delta configurations have all three windings connected in series as shown in Figure 1-13. The phase leads are connected to the three common points where windings are joined.

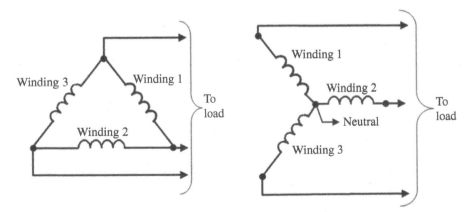

Figure 1-13   Delta and wye configurations.

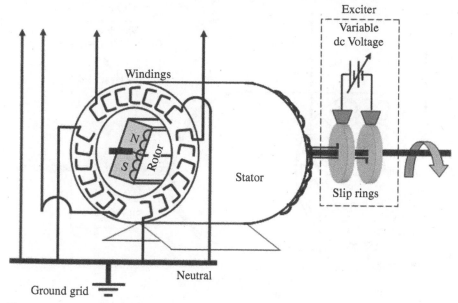

Figure 1-14   wye-connected generator.

Wye

The wye configuration connects one lead from each winding to form a common point called the *neutral*. The other three phase leads are brought out of the generator separately for external system connections. The neutral is often *grounded* to the station ground grid for voltage reference and stability purposes. Grounding the neutral is discussed later.

Wye and Delta Stator Connections

Electric power plant generators use either wye or delta stator connections. The phase leads from the generator are connected to the plant's *step-up transformer* (not shown yet), where the generator output voltage is increased significantly to transmission voltage levels for the efficient transportation of electrical energy. Step-up transformers are discussed later in this book. Figures 1-14 and 1-15 show both the wye and the delta generator connections.

THREE TYPES OF ELECTRICAL LOAD

Devices that are connected to the power system are referred to as electrical *load*. A toaster, refrigerator, bug zapper, etc. are considered electrical load. There are three types of electrical load. They vary according to their *leading* or *lagging* time or phase relationship between voltage and current.

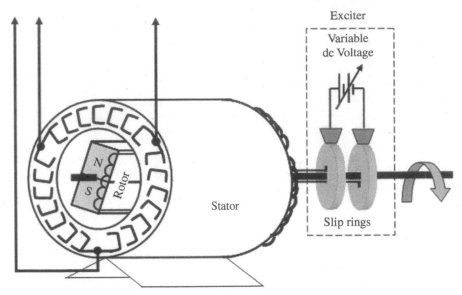

Figure 1-15   Delta-connected generator.

The three load types are *resistive, inductive, and capacitive*. Each type has specific characteristics that make them unique from each other. Understanding the differences between these load types help explain how power systems can operate efficiently. Power system engineers, system operators, maintenance personnel, and others try to maximize system efficiency on a continuous basis by having a good understanding of the three types of loads and how they interact with each other. They understand how having them work together efficiently can minimize system losses, provide additional equipment capacity, and maximize system reliability.

The three different types of load; Resistive, Inductive, and Capacitive are summarized below. The standard units of measurement are in parentheses and their symbols and abbreviations follow.

Resistive Load

The resistance in a wire (i.e., conductor) causes friction and reduces the amount of current flow if the voltage remains constant. By-products of this electrical friction are heat and light. The unit of resistance is the *Ohm*, named after George Ohm, a German mathematician and physicist. The unit of electrical power associated with resistive load is *Watts*. Examples of resistive load are shown in Figure 1-16; Light bulbs, toasters, electric hot water heaters, etc. are resistive loads.

Resistive
(Ohms)

R

Figure 1-16   Resistive loads.

Inductive
(Henrys)   L

Figure 1-17   Inductive loads.

Inductive Load

Inductive loads require a magnetic field to operate. All electrical loads that have a coil of wire to produce the magnetic field in order to function are called inductive loads. Examples of inductive loads are shown in Figure 1-17; hair dryers, fans, blenders, vacuum cleaners, drills, and many other motorized devices. In essence, all motors are inductive loads. The unique difference about inductive load, as compared to the other load types, is that the current in an inductive load *lags* the applied voltage. Inductive loads take time to develop their magnetic field when the voltage is applied, so the current is delayed. The unit of inductance is the *Henry*, named after Joseph Henry, a U.S. physicist.

Regarding electrical motors, load placed on the spinning shaft to perform work functions draws what is referred to as *real* power (i.e., Watts) from the electrical energy source. In addition to real power, what is referred to as *reactive* power is also drawn from the electrical energy source to produce the magnetic fields in the motor. The *total power* consumed by the motor is therefore the sum of both real and reactive power. The units of electrical power associated with reactive power are called *positive VARs*. (The acronym VARs stands for volts-amps-reactive.)

Capacitive Load (Figure 1-18)

A capacitor is a device made of two metal conductors separated by an insulator, called a *dielectric* (i.e., air, paper, glass, and other non-conductive materials). These dielectric materials become charged when voltage is applied to the attached conductors. Capacitors can remain charged long after the voltage source has been removed. Examples of capacitor loads are old TV picture tubes, long extension cords, discrete components used in electronic devices, and many other devices.

Opposite to inductors, the current associated with capacitors *lead* (instead of lag) the voltage because of the time it takes the dielectric material to charge up to full voltage from the charging current. Therefore, it is said that the current in a capacitor leads the voltage. The unit of capacitance is called *Farad*, named after Michael Faraday, a British physicist.

Capacitive
(Farads)

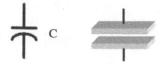

Figure 1-18   Capacitive loads.

Similar to inductors, the power associated with capacitors is also called reactive power, but has the opposite polarity. Thus, inductors have positive VARs and capacitors have *negative VARs*. Note: the negative VARs of inductors can be cancelled by the positive VARS of capacitors to have a net zero reactive power requirement. Therefore, when VARs cancel total power equals real power. How capacitors cancel out inductors in electrical circuits to improve system efficiency will be discussed later.

As a general rule, capacitive loads are not items that people purchase at the store in massive quantities like they do resistive and inductive loads. For that reason, power companies install capacitors on their power systems on a regular basis to maintain a reactive power balance with the typically high inductive demand.

Note: it is very helpful to understand how the three types of load (i.e., resistors, inductors, and capacitors) interact in power systems because their relationships influence system load, overall losses, revenues, and system reliability. These load types are discussed in more detail later in this book.

CHAPTER 2

GENERATION

CHAPTER OBJECTIVES

After completing this chapter, the reader will be able to:

- ☑ *Explain what is meant by "Real-Time Generations"*
- ☑ *Discuss the operation of several different types of generation plants (i.e., steam, nuclear, wind, etc.)*
- ☑ *Describe the environmental considerations for the different power plant types*
- ☑ *Describe the growth statistics associated with renewable energy resources*

REAL-TIME GENERATION

Power plants produce electrical energy on a *real-time* basis. Electric power systems do not store energy such as most gas or water systems. For example, when a toaster is switched on and consuming electrical energy from the system, the associated generating plants immediately see this as new load addition and slightly slow down. As more and more load (i.e., toasters, lights, motors, etc.) are switched on, generation output and prime mover rotational shaft energy must be increased to balance the load demand on the system. Unlike water utility systems that store water in tanks located up high on hills or tall structures to serve real-time demand, electric power systems must control generation in real time to balance load on demand. Water is pumped into the tank when the tank is low, allowing the pumps to turn off during low and high demand periods. Electrical generation always produces electricity on an "as needed" basis. Note, some generation units can be taken offline during light load conditions, but there must always be enough generation online to maintain frequency and voltage during light and heavy load conditions.

There are electrical energy storage systems such as batteries, but, electricity found in interconnected ac power systems are real-time energy supply systems and not energy storage systems. The challenge is to provide energy storage capacity when demand calls for it. Pumped storage systems (discussed in more detail later) provide that bulk scale electrical energy storage. Battery energy, a dc device, must be converted to ac when augmenting power system demand. This opens the door for battery research or better use of existing battery systems. For example, imagine everyone

Electric Power System Basics for the Nonelectrical Professional, Second Edition. Steven W. Blume.
© 2017 by The Institute of Electrical and Electronics Engineers, Inc. Published 2017 by John Wiley & Sons, Inc.

having electric cars and rooftop solar voltaic systems; the aggregate vehicle battery energy could be a potential storage system while the solar dc/ac inverters are used in conjunction with grid-tie systems.

POWER PLANTS AND PRIME MOVERS

Power generation plants produce the electrical energy that is ultimately delivered to consumers through transmission lines, substations, and distribution lines. Generation plants or power plants consist of three-phase generator(s), the *prime mover*, the energy source to serve as the generator's prime mover, control room, and substation. The generator portion has been discussed already. The prime movers and their associated energy sources are the focus of this section.

The mechanical means of turning the generator's rotor is called the prime mover. The prime mover's energy sources include the conversion process from raw fuel, such as coal to the end product being steam that will turn the turbine. The bulk of electrical energy produced in today's interconnected power systems for prime movers is normally produced through a conversion process from coal, oil, natural gas, and nuclear for steam-driven turbines and water for hydro turbines. To a lesser degree, electrical power is produced by wind, solar, geothermal, and biopower energy resources.

The more common types of energy resources used to generate electricity and their associated prime movers that are discussed in this chapter include:

- Steam Turbines
    - Fossil Fuels (coal, gas, oil)
    - Nuclear
    - Geothermal
    - Solar Heated Steam
- Hydro Turbines
    - Dam and River
    - Pump Storage
- Combustion Turbines
    - Diesel
    - Natural Gas
    - Combined Cycle
- Renewable Energy
    - Wind Turbines
    - Solar Direct (Photovoltaic)
    - Biopower (Wood and Agricultural Residues)

Steam Turbine Power Plants

High-pressure and high-temperature steam is created in a boiler, furnace, or heat exchanger and moved through a *steam turbine generator* (STG) which converts the steam's energy into rotational energy into a shaft, which turns the generator's rotor. The steam turbine's rotating shaft is directly coupled to the generator rotor. The STG shaft speed is tightly controlled by steam throttle valves and governors for it is directly related to the frequency of electrical power being produced.

High-temperature, high-pressure steam is used to turn steam turbines which ultimately turn the generator rotor. Temperatures in the order of 1000°F and pressures in the order of 2000 pounds per square inch (psi) are commonly used in large steam power plants. Steam at this pressure and temperature is called *superheated steam*. This condition of steam is sometimes referred to as *dry steam*.

The steam's pressure and temperature drop significantly after it is applied across the *first stage* turbine blades. Turbine blades make up the fan-shaped rotor where steam is directed, thus turning the shaft. The superheated steam is reduced in pressure and temperature after it passes through the turbine. The reduced steam can be routed through a *second stage* set of turbine blades where additional steam energy is transferred to the turbine shaft. This second stage equipment is significantly larger than the first stage to allow for additional expansion and energy transformation. In some power plants the steam following the first stage is redirected back to the boiler where it is re-heated and then sent back to the second turbine stage for a more efficient energy transformation.

Once the energy of the steam has been transferred to the turbine shaft the low-temperature and low-pressure steam has basically exhausted its energy and must be fully *condensed* back to water before it can be recycled. The condensing process of steam back to water is accomplished by a *condenser* and/or *cooling tower(s)*. Once the used steam is condensed back to warm water, the *boiler feed pump* (BFP) pumps the warm water back to the boiler and recycled into superheated steam. This is a closed loop processes. Some water has to be added in the process due to small leaks and evaporation.

The condenser takes cold water from nearby lakes, ponds, rivers, oceans, deep wells, cooling towers, and other water sources and pumps it through pipes in the condenser. The used steam passes by the relatively cold water pipes of the condenser and causes dripping to occur. The droplets are collected at the base of the condenser (the well) and pumped back to the boiler by the BFP.

The overall steam generation plant efficiency in converting fuel heat energy into mechanical rotation energy and then into electrical energy ranges 25–35%. Although a relatively low efficiency system, steam turbine generation is very reliable and is commonly used as base load generation units in large electric power systems. Most of the inefficiency in steam turbine generation plants comes from the loss of heat into the atmosphere in the furnace/boiler process.

Fossil Fuel Power Plants

Fossil fuel power plants use steam turbines that burn coal, oil, natural gas, or just about any combustible material as the fuel resource. However, each fuel type requires a unique set of accessory equipment to inject fuel into the boiler, control the burning process, vent and exhaust gases, capture unwanted by-products, etc.

Some fossil fuel power plants can switch fuels. For example, it is common for an oil plant to convert to natural gas when gas prices are less expensive than oil. Most of the time it is not practical to convert a coal burning power plant to oil or gas unless it has been designed for conversion. The processes are usually different enough to not be cost effective.

Coal is burned two different ways in coal-fired plants. First, in traditional coal-fired plants, the coal is placed on metal conveyor belts inside the furnace/boiler chamber. The coal is burned while on the belt and as the belt slowly traverses the bottom of the boiler. Ash falls through the chain conveyor belt and is collected below where it is usually sold as a useful by-product in other industries.

In pulverized coal power plants, the coal is crushed into a fine powder and injected into the furnace where it is burned similar to a gas. Pulverized coal is mixed with air and ignited in the furnace. Combustion by-products include solid residue (ash) that is collected at the bottom of the furnace and gases that include fine ash, NO_2, CO, and SO_2 that are emitted into the atmosphere through the stack. Depending on local environmental regulations, scrubber and bag house equipment may be required and installed to collect most of these by-products before they reach the atmosphere. Scrubbers are used to collect the undesirable gases to improve the quality of the stack output emissions. Bag houses are commonly used to help collect fly ash.

Some of the drawbacks that could be encountered with coal-fired steam generating power plants are:

- Environmental concerns from burning coal (i.e., acid rain)
- Transportation issues regarding rail systems for coal delivery
- Distance of transmission lines to remote power plant locations

Figure 2-1 shows the layout of a typical coal-fired steam power plant. Notice the steam line used to transfer superheated steam from the boiler to the turbine and then through the condenser where it is returned to a water state and recycled. Notice the steam turbine connected to the generator. The turbine speed is controlled by the amount of applied steam to control frequency. When load picks up on the electrical system, the turbine shaft speed slows down and more steam is then placed on

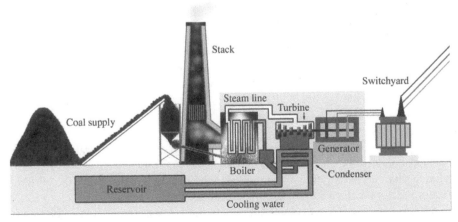

Figure 2-1   Steam power plant.

Figure 2-2   Coal power plant. Reproduced with permission of Fotosearch.

the turbine blades to maintain frequency. Notice how coal is delivered to the boiler and burned. Exhaust is vented through the stack. Scrubber and bags remove the by-products before they enter the atmosphere. Water from a nearby reservoir is pumped to the condenser where it is used to convert steam back into water and recycled.

Figure 2-2 is an actual coal-fired steam turbine power plant. The ramp in front lifts the coal to the pulverizer where it is crushed before injection into the furnace/boiler and burned. Plant operators must be careful to not allow spontaneous combustion of coal to occur while stored in the yard.

Nuclear Power Plants

Nuclear power plants, such as the one shown in Figure 2-3, use a controlled nuclear reaction to make heat in order to produce steam needed to drive an STG.

All nuclear plants in the United States must conform to the Nuclear Regulatory Commission's rules and regulations. Extensive documentation is required to establish that the proposed design can be operated safely without undue risk to the public. Once the Nuclear Regulatory Commission issues a license, the license holder must maintain the license and the reactor in accordance with strict rules, usually called Tech specs. Compliance to these rules and regulations in conjunction with site inspections insures that a safe nuclear power plant is in operation.

Nuclear Energy

Atoms are the building blocks from which all matter is formed. Everything is made up of atoms. Atoms are made up of a nucleus (with protons and neutrons) and orbiting

Figure 2-3   Nuclear power plant. Reproduced with permission of Fotosearch.

electrons. The number of atomic particles (i.e., sum of neutrons, protons, and electrons) determines the atomic weight of the atom and type of elements on the periodic table. Nuclear energy is contained within the center of such atoms (i.e., nucleus) where the atom's protons and neutrons exist. Nature holds the particles within the atom's nucleus together by a very strong force. If a large nucleus element (such as Uranium 235) is split apart into multiple nuclei of different element compositions, generous amounts of energy are released in the process. The heat emitted during this process (i.e., *nuclear reaction*) is used to produce steam energy to drive a turbine generator. This is the foundation of a nuclear power plant.

There are basically two methods used to produce nuclear energy in order to produce heat to make steam. The first process is called *fission*. Fission is the splitting of large nuclei atoms such as Uranium inside a nuclear reactor to release energy in the form of heat to be used to produce steam to drive steam turbine electrical power generators. The second process is called *fusion*. Fusion is the combining of small nuclei atoms into larger ones resulting in an accompanying release of energy (heat). However, fusion reactors are not yet used to produce electrical power because it is difficult to overcome the natural mutual repulsion force of the positively charged protons in the nuclei of the atoms being combined.

In the fission process certain heavy elements, such as Uranium, are split when a neutron strikes them. When they split, they release energy in the form of *kinetic energy* (heat) and *radiation*. Radiation is subatomic particles or high-energy light waves emitted by unstable nuclei. The process not only produces energy and radiation, but also provides additional neutrons that can be used to fission other Uranium nuclei and in essence start a chain reaction. The controlled release of this nuclear energy using commercial grade fuels is the basis of electric power generation. The uncontrolled release of this nuclear energy using more highly enriched fuels is the basis for atomic bombs.

The reactor is contained inside an obvious *containment shell*. It is made up of extremely heavy concrete and dense steel in order to minimize the possibility of a reactor breech due to accidental situations. Nuclear power plants also have an emergency backup scheme of injecting *boron* into the reactor coolant. Boron is an element that absorbs neutrons very readily. By absorbing neutrons, the neutrons are not available to continue the nuclear reaction, and the reactor shuts down.

The most widely used design for nuclear reactors consists of a heavy steel pressure vessel surrounding the *reactor core*. The reactor core contains the Uranium fuel. The fuel is formed into cylindrical ceramic pellets about half an inch in diameter, which are sealed in long metal tubes called *fuel tubes*. The tubes are arranged in groups to make a *fuel assembly*. A group of fuel assemblies forms the *reactor core*.

Controlling the heat production in nuclear reactors is accomplished by using materials that absorb neutrons. These control materials or elements are placed among the fuel assemblies. When the control elements, or *control rods* (as they are often called), are pulled out of the core, more neutrons are available and the chain reaction increases, producing more heat. When the control rods are inserted into the core, more neutrons are absorbed, and the chain reaction slows down or stops, producing no heat altogether. The *control rod drive system* controls the actual output power of the electric power plant.

Most commercial nuclear reactors use ordinary water to remove the heat created by the fission process. These are called *light water reactors*. The water also serves to slow down, or *moderate* the neutrons in the fission process. In this type of reactor, control mechanisms are used such that the chain reaction will not occur without the water to serve as a moderator. In the United States, there are two different types of light water reactor designs used, the *pressurized water reactor* (PWR) and the *boiling water reactor* (BWR). Note there are approximately 70% more PWR nuclear power plants in the United States than BWR.

Pressurized Water Reactor (PWR)

The basic design of a pressurized water reactor is shown in Figure 2-4. The reactor and the primary steam generator are housed inside a containment structure. The structure is designed to withstand events such as small airplane crashes. The PWR steam generator separates the radioactive water that exists inside the reactor from the steam that is going to the turbine outside the shell.

In a PWR, the heat is removed from the reactor by water flowing in a closed pressurized loop. The heat is transferred to a second water loop through a *heat exchanger* (or *steam generator*). The second loop is kept at a lower pressure, allowing the water to boil and create steam, which is used to turn the turbine generator and produce electricity. Afterward, the steam is condensed back into water and returned to the heat exchanger where it is recycled into useable steam.

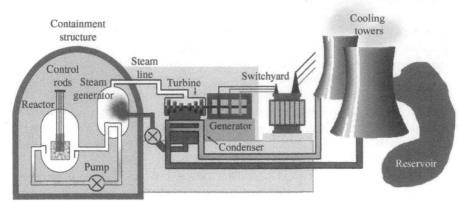

Figure 2-4   Pressurized water reactor.

The normal control of the reactor power output is by means of the *control rod system*. These control rods are normally inserted and controlled from the top of the reactor. Because the control rods are inserted and controlled from the top of the reactor, the design also includes special springs and release mechanisms so that if all power is lost, the control rod will be dropped into the reactor core by gravity to shut down the reactor.

Advantages and Disadvantages of PWR   As with any design, there are advantages and disadvantages of PWRs. A major design advantage is the fact that fuel leaks, such as ruptured fuel rods, are isolated to the core and primary loop. That is, radioactive material contained inside the fuel is not allowed to go outside of the containment shell. The PWR can be operated at higher temperature/pressure combinations, and this allows an increase in the efficiency of the turbine generator system.

Another advantage is that it is believed that a PWR is more stable than other designs. This is because boiling is not allowed to take place inside the reactor vessel, and therefore the density of the water in the reactor core is more constant. By reducing the variability of the water density, controls are somewhat simplified.

The biggest disadvantage appears to be the fact that the reactor design is more complicated. It is necessary to design for extremely high pressures and temperatures in order to ensure that boiling does not take place inside the reactor core. The use of high-pressure vessels makes the overall reactor somewhat more costly to build. Finally, under certain circumstances, the PWR can produce power at a faster rate than the cooling water can remove heat or condense the water. If this event takes place, there is a high probability of fuel rod damage.

Boiling Water Reactor (BWR)

Figure 2-5 shows the boiling water reactor (BWR). Again there is a reactor building or containment shell where the nuclear reactor and some of its complement equipment are located. The reactor housing of the BWR tends to be larger than the PWR and looks almost like an inverted light bulb.

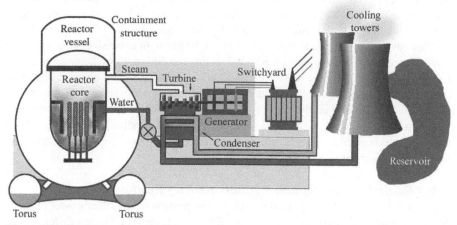

Figure 2-5 Boiling water reactor.

In a BWR, water boils inside the reactor itself, and the steam goes directly to the turbine generator to produce electricity. Similar to other steam power plants, the steam is condensed and reused. Note that the turbine building is closely coupled to the reactor building, and special constraints exist in entering the turbine building because the water can pick up radioactivity.

Note the *Torus* at the bottom of the reactor. If there should be a reactor rupture, the water inside the reactor will flash into steam and create a very high pressure surge on the reactor building. The reactor Taurus is filled with cold water, which will instantly condense the steam. The Torus system ensures that the pressure inside the containment dome never exceeds an acceptable level.

As with the PWR, the reactor housing contains the fuel core and supplies to water flow paths. The reactor recirculation system consists of the pumps and pipes that circulate the water through the reactor. The water circulating through the reactor actually goes into the turbine itself and then condensed water goes back into the reactor. The steam separator in the reactor shell separates the water from the steam and allows the steam to pass to the steam generator. The separated water is returned to the reactor for recirculation.

The BWR utilizes one cooling loop. Both water and steam exist in the reactor core (i.e., a definition of boiling). Reactor power is controlled by positioning the control rods from start up to approximately 70% of rated power. From 70% to 100% of rated power, the reactor power is controlled by changing the flow of water through the core. As more water is pumped through the core and more steam generated, more power is produced. In the BWR, control rods are normally inserted from the bottom. The top of the reactor vessel is used to separate water and steam.

Advantages and Disadvantages of BWR   A major advantage of the BWR is that the overall thermal efficiency is greater than that of a PWR because there is no separate steam generator or heat exchanger. Controlling the reactor is a little easier than a PWR because it is accomplished by controlling the flow of water through the core. Increasing the water flow increases the power generated. Because of the nature of the design, the reactor vessel is subjected to less radiation, and this is considered to be an advantage because some steels become brittle with excessive radiation.

The greatest disadvantage of the BWR is that the design is much more complex. It requires a larger pressure vessel than the PWR because of the amount of steam that can be released during an accident. This larger pressure vessel also increases the cost of the BWR. Finally, the design does allow a small amount of radioactive contamination to get into the turbine system. This modest radioactivity requires that anybody working on the turbine must wear appropriate protective clothing and use the proper equipment.

Other Related Topics (Optional Supplementary Reading)

The overall function or design of the non-nuclear portion of a nuclear power plant is on the order of complexity as is fossil-fueled power plants. The biggest difference is the degree of documentation that must be maintained and submitted to the regulatory authorities for proof that the design and operation is safe. Roughly speaking, there are about 80 separate systems in the nuclear power plant. The systems that are most critical are those which control the power and/or limit the power output of the plant.

Environmental   One of the greatest advantages of a nuclear plant, especially with today's concerns about global warming and generation of carbon dioxide due to burning, is the fact that a nuclear plant essentially adds zero emissions to the atmosphere. There is no smoke stack!

SCRAM   A reactor *SCRAM* is an emergency shutdown situation. Basically, all control rods are driven into the reactor core as rapidly as possible to shut down the reactor to stop heat production. A SCRAM occurs when some protective device or sensor signals the control rod drive system. Some typical protective signals that might initiate or trigger a SCRAM include a sudden change in neutron production, a sudden change in temperatures inside the reactor shell, sudden change in pressures or other potential system malfunction.

By inserting the control rods into the reactor core, the reactor power is slowed down and/or stopped because the control rod materials absorb neutrons. If the neutrons are absorbed, they cannot cause fission in additional uranium atoms.

Anytime there is a reactor SCRAM, the cause must be fully identified and appropriate remedial actions taken before the reactor can be restarted. Needless to say, a reactor SCRAM usually results in a great deal of paperwork to establish the fact that the reactor can be safely restarted.

There are various theories as to where the term SCRAM came from. One theory says that around the World War II era the original nuclear reactors were controlled manually. As a safety measure, the reactor was designed so that control rods would drop by gravity into the reactor core and absorb the neutrons. The control rods were held up by a rope. In case of emergency, the rope was to be cut to allow the rods to drop. The person responsible for cutting the rope in case of any emergency was called the SCRAM. According to the Nuclear Regulatory Commission, SCRAM stands for *"safety control rod axe man."* Now, SCRAM stands for any emergency shutdown of the reactor for any reason.

   Equipment Vibration   Equipment vibration is probably the biggest single problem in nuclear power plants. Every individual component is monitored by a central computer system for vibration indications. If excessive vibration is detected the system involved must be quickly shutdown. (Note: this is also true of regular steam plants. If excessive vibration is detected in the turbine or generator they will be shut down.)
   Nuclear power plants seem to be particularly susceptible to vibration problems, especially on the protective relay panels. Excessive vibration can cause inadvertent relay operations, shutting down a system or the complete plant.
   Microprocessor-based protection relay equipment is basically immune to vibration problems, but there is a perception that the solid-state circuits used in such relays may be damaged by radiation. Most nuclear power plants still use electromechanical relays as backup to the microprocessor solid-state relays.

Hydroelectric Power Plants

Hydro power plants capture the energy of moving water. There are multiple ways hydro energy can be extracted. Falling water such as in a penstock, flume, or water wheel can be used to drive a hydro turbine. Hydro energy can be extracted from flowing water such as the lower section of dams where the pressure forces water flow. Hydroelectric power generation is efficient, cost effective, and environmentally cooperative. Hydro power production is considered to be a renewable energy source because the water cycle is continuous and constantly recharged.
   Water flows much slower through a hydro turbine than high-pressure steam turbine. Therefore, several rotor magnetic poles are used to reduce the rotational speed requirement of the hydro turbine shaft.
   Hydro units have a number of excellent advantages. The hydro unit can be started very quickly, and brought up to full-load in a matter of minutes. In most cases, little or no startup power is required. A hydro plant is almost by definition, a *black start* unit. Black start means that electrical power is not needed first in order to start a hydro power plant. Hydro plants have a relatively long life; 50–60 year life spans are common. Some hydroelectric power plants along the Truckee river in California have been in operation for well over 100 years. Figure 2-6 shows a typical hydroelectric power plant.
   The cross section of a typical low head hydro installation is shown in Figure 2-7. Basically, the water behind the dam is transported to the turbine by means of a *penstock*. The turbine causes the generator to rotate producing electricity, which is then delivered to the load center over long-distance, high-voltage power lines. The water coming out of the turbine goes into the river.

Pumped Storage Hydro Power Plants

Pumped storage hydro power production is a means of actually saving electricity for future use. Power is generated from water falling from a higher lake to a lower lake

Figure 2-6   Hydroelectric power. Reproduced with permission of Photovault.

during peak load periods. The operation is reversed during off-peak conditions by pumping the water from the lower lake back to the upper lake. A power company can obtain high value power during peak load generation periods by paying the lower cost to pump the water back during off-peak periods. Basically, the machine at the lower level is reversible; hence it operates as a hydro-generator unit or a motor-pump unit.

One of the problems associated with pumped storage units is the process of getting the pumping motor started. Starting the pumping motor using the system's power line would usually put a low voltage sag condition on the power system. The voltage sag or dip could actually cause power quality problems. In some cases, two turbines are used in a pumped storage installation. One of the turbines is used as a generator to start the other turbine as a pump. Once the turbine is turning, the impact on the power system is much less, and the second turbine can then be started as a motor-pump.

Figure 2-8 shows a cross-sectional view of the Tennessee Valley Authority's pumped storage plant at Raccoon Mountain. The main access tunnel was originally used to bring all of the equipment into the powerhouse: the turbine, the pumps, and the auxiliary equipment. Note that Tennessee Valley Authority installed a visitor center at the top of the mountain so that the installation could be viewed by the general public.

Combustion Turbine Generation Plants

Combustion turbine (CT) power plants burn fuel in a jet engine and use the hot exhaust gases to spin an STG. In the combustion turbine generator, air is compressed to a very high pressure, fuel is then injected into the compressed air and ignited, producing high-pressure and high-temperature exhaust gases. The exhaust is moved

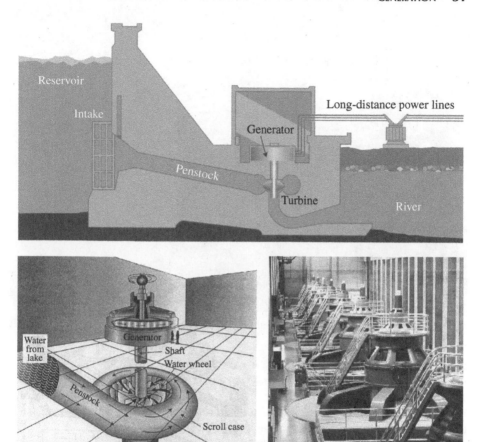

Figure 2-7   Hydro power plant.

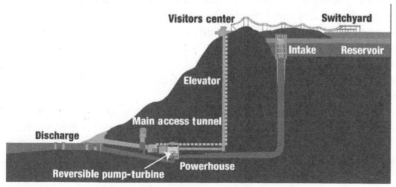

Figure 2-8   Pump storage power plant.

Figure 2-9   Combustion turbine power plant.

though turbine blades much the same way steam is moved through turbine blades in a steam power plant. The exhaust gas movement through the combustion turbine results in the rotation of the generator rotor, thus producing electricity. The exhaust from the CT remains at a very high temperature and pressure after leaving the turbine. Figure 2-9 shows a combustion turbine generator.

One of the advantages of combustion turbines is that they can actually be designed to be remotely controlled for unmanned sites. They offer fast start-up times and fast installation times. In some cases, the purchase of the combustion turbine generator system can be "turn-key," that is, the owner simply contracts for a complete installation and takes over when the plant is finished and ready to operate. In most cases, the combustion turbine generator package is a completely self-contained unit. In fact, some of the smaller capacity systems are actually built on trailers so that they can be moved quickly to sites requiring emergency generation.

Combustion turbines can be extremely responsive to power system changes. They can go from no-load to full-load and vice versa in a matter of seconds or in a matter of minutes.

The disadvantages are limited fuel options (i.e., diesel fuel, jet fuel, or natural gas) and inefficient use of exhaust heat.

There are several environmental issues related to the use of combustion turbines. Without appropriate treatment, the exhaust emissions can be very high in undesirable gases. The high temperatures in the combustion chamber will increase the production of nitric oxide gases and their emissions. Depending on the fuel used, there can be particulate emission problems. That is, particles or other materials tend to increase the opacity (i.e., smoke) of the gases. Sound levels around combustion turbine installations can be very high. Special sound reduction systems are available and used. (Note, combustion turbines are typically jet engines very similar to those heard at airports.)

The heat rate or efficiency of a simple cycle combustion turbine is not very good. The efficiencies are somewhere in the range of 200–40% maximum.

One effective way to overcome some of the cost is to incorporate a heat exchanger to the exhaust gases to generate steam that will drive a secondary steam turbine. Many CTs are used as "combined cycle power plants."

Combined Cycle Power Plants (Combustion and Steam)

The *combined cycle power plant* consists of two means of generation; combustion turbine and steam turbine. The combustion turbine is similar to a jet engine whose high temperature and high pressure exhaust spins a turbine whose shaft is connected to a generator. The hot exhaust is then coupled through a *heat recovery steam generator* (HRSG) that is used to heat water, thus producing steam to drive a secondary STG. The combustion turbine typically uses natural gas as the fuel to drive the turbine blades.

The advantage of a combined cycle (CC) system is that in addition to the electrical energy produced by the fuel combustion engine, the exhaust from the engine also produces electrical energy. Another potential benefit of CC plants is the end user can have steam made available to assist other functions such as building heat, hot water, and production processes that require steam (such as nearby paper mills). Therefore for one source of fuel (i.e., natural gas), many energy services are provided (electrical energy, steam, hot water, and building heat). Some CC can reach efficiencies near 90%. Figure 2-10 shows a combined cycle power plant.

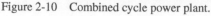

Figure 2-10   Combined cycle power plant.

Renewable Energy

Renewable energy is primarily made up of the following generation types:

- Wind
- Solar photovoltaic (PV)
- Concentrating solar power (CSP)
- Geothermal
- Biopower
- Hydropower (already discussed)

The chart in Figure 2-11 shows the global growth in renewable energy by generation type according to the National Renewable Energy Association (NREL):

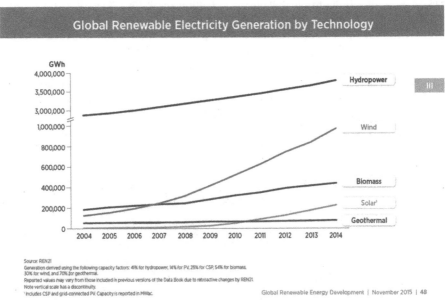

Figure 2-11   Global renewable energy. Reproduced with permission of National Renewable Energy Laboratory.

Some of these renewable energy resource types have idiosyncrasies that could tend to constrain growth. For example, hydro growth is limited due to the fact that most rivers in the United States already have fully implemented hydro generation. Geothermal has limited growth potential because there are limited areas in the United States that can support this type of generation. Biopower (i.e., biomass), especially the agricultural residue type requires more research and creative production techniques to make this resource a significant contributor in the USA's energy demand. This leaves wind and both types of solar (concentrated-mirrors and photovoltaic) to represent large-scale growth opportunities in the United States. As we will discuss later, the operational constraints from having too much wind or solar on our nations power grids could limit growth in these resources. There is still room, however, to accommodate more growth in wind and solar generation. The caution comes into play when large amounts of existing coal generation and other fossil fuel plants retire and are replaced with wind and solar. This could impact the reliability of the nation's power grid. As we will discuss later, spinning inertia is critical in maintaining system reliability and both solar and wind generation can affect this requirement if too much is used to power our grids.

Wind Turbine Generators

Wind generation has increased in popularity and technology has improved tremendously over the last decade. In the year 2006, the total installed capacity of US wind generation was about 11,000 MW. In 2014, the amount of installed wind capacity in the United States is 181,700 MW (NREL). On and offshore wind turbine generators are continuing to be installed worldwide. The total installed capacity worldwide in 2014 is about 318,800 MW and the average turbine size is 1.94 MW (Figure 2-12 shows typical wind generators.

Although wind turbine generators tend to have high cost per kWh produced, wind energy prices are declining and it is increasingly becoming a competitive choice of power. With improving technology and siting techniques, wind energy is increasingly becoming one of the most affordable forms of electricity today.

There is a concern about the availability of wind on a constant basis. Most power companies do not consider wind generators to be *base load* units. Base load implies readily available and that they become part of the 24-hour generation production schedule. They are brought online when available.

Figure 2-12   Wind power. Reproduced with permission of Fotosearch.

Figure 2-13   Offshore wind turbines. Courtesy of ABB.

Basically, the concept of wind power is that the wind energy is converted into electrical energy by means of modern wind turbines. One interesting characteristic of wind power is the fact that power produced is proportional to the cube of the wind speed. In other words, if the wind speed is doubled, the power produced is tripled or increased by a factor of eight. Thus, what might appear to humans as modest fluctuations in wind speed, or breezes, can severely impact wind power production and system frequency stability.

Offshore wind generation like that shown in Figure 2-13 has very favorable attributes when it comes to wind power efficiency and consistency. Wind power efficiency is rated at sea level and decreases as elevation increases due to air density being much lower at high elevations. The higher the air density (sea level) the higher the power output for the same wind speed. However, offshore wind does have its visual impact controversies.

Installation of wind power generators require selecting sites that are relatively unrestricted to wind flow, have high air density (sea level) and within close proximity to suitable power lines.

Wind power is considered a renewable energy source since wind sustains itself. Wind power is accepted as free energy with no fuel costs. There is however a limit to how much wind power can be added to the grid due to required system stability and reliability constraints. This concern is discussed later in this book.

Solar Reflective Power

Solar energy comes in two general categories; solar reflective (mirrors) and solar photovoltaic (panels). Both forms of solar energy production are environmentally friendly as they produce no pollution and considered renewable energy resources. Large-scale *solar reflective power plants* (also called "concentrating solar power" or

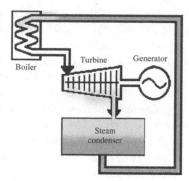

Figure 2-14   Reflective solar power plant. Reproduced with permission of Fotosearch.

(CSP) require a substantial amount of land area as well as specific orientation with the sun to capture the maximum energy possible with high efficiency.

Solar energy is reflected off mirrors and concentrated on a centralized boiler system. The mirrors are parabolic-shaped and motorized to focus the sun's energy toward the receiver tubes in the collector area of the elevated boiler. The receiver tubes contain a heat transfer fluid used in the steam–boiler–turbine system. The collector area housing the receiver tubes absorbs the focused sun energy to gain 30–100 times normal solar energy. The fluid in these tubes can reach operating temperatures in excess of 400°C. The steam drives the turbine and then goes through a condenser for conversion back to liquid before being reheated in the boiler system. A typical solar reflective power plant is shown in Figure 2-14.

This form of solar power has grown significantly over the last decade. The US CSP capacity in 2014 has grown to 1685 MW.

Solar Direct Generation (Photovoltaic)

The *photovoltaic* (sometimes called "voltaic" or "PV" for short) type solar power plant converts the sun's energy directly into electrical energy. A photovoltaic array is shown in Figure 2-15. This type of production uses various types of films or

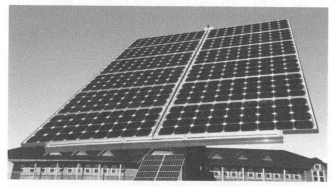

Figure 2-15   Direct solar photovoltaic. Reproduced with permission of Fotosearch.

special materials that convert sunlight into direct current (dc) electrical energy systems. Individual panels are connected in series and parallel to obtain the desired dc output voltage and current ratings. This dc energy is then converted to utility ac energy by means of a device called an *inverter*. The ac power from the inverter is connected to the power grid. Some inverter systems use an energy storage device (i.e., battery) to provide electrical power during off sun peak periods.

One can just imagine how the future brings many exciting opportunities for residential customers that have solar PV grid-tie systems, electric vehicles, low-voltage lighting, home energy management systems, and other new control and automation technology devices. Many consumers are already employing apps and digital devices that are controllable through smart phones and other devices.

The photovoltaic solar cell is small in size, typically made of 1.5 Vdc solar cells capable of producing approximately 20 mA of electrical current each. A typical solar photovoltaic panel that is made up of several small solar cells that measure 4 feet tall by 2 feet wide would produce approximately 250 W of electrical power. Therefore, twenty of these 4 × 2 feet panel would supply 5 kW of power each sunlight hour, capable of running several household appliances. And, from an energy standpoint; if there are 5 sunlight hours per day for 30 days (1 month), the energy produced from this solar system would be approximately 750 kWh. This is a typical residential application. Then it is a matter of what state and federal incentives are available to make this a viable option. One has to determine if there are enough federal tax credits, special bank loan rates, attractive utility grid-tie tariff agreements available and other encouraging incentives that make a residential solar project like this affordable and worthwhile. Already this approach to augment utility power is in high pursuit and continues to grow.

Utilities and special interest groups are building or have built several large-scale solar farms to help meet growing demand for electrical energy. This environmentally safe approach to electrical energy production is growing fast. Note, similar to wind generation, there are some drawbacks to having too much solar power on the grid and this will be discussed in more detail later in this book under grid reliability.

According to the NREL, solar growth has been substantial. The chart in Figure 2-16 shows the growth in installed capacity and energy generation in the United States.

The chart in Figure 2-17 shows utility and residential growth in solar power over a recent 5-year period according to NREL.

Solar power is environmentally friendly as it produces no pollution. Solar power technology, installed capacity, utility operations, governmental incentives, and expansion into new applications continue to progress positively as we head into the future, balancing electrical generation and consumption.

Geothermal Power Plants

Geothermal power plants use hot water and/or steam located underground to produce electrical energy. The hot water and/or steam are brought to the surface where heat exchangers are used to produce clean steam in a secondary system for use with turbines. Clean steam has no sediment growth inside pipes and other equipment

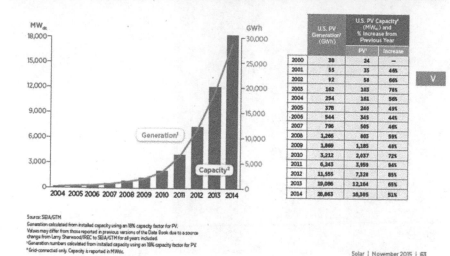

Figure 2-16   Solar installation growth. Reproduced with permission of National Renewable Energy Laboratory.

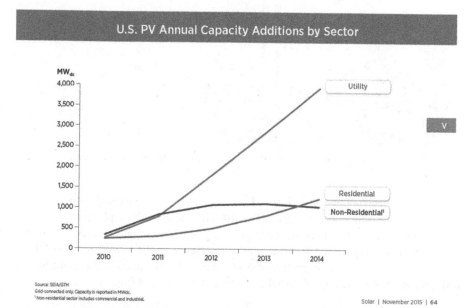

Figure 2-17   Solar growth by sector. Reproduced with permission of National Renewable Energy Laboratory.

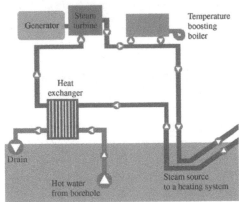

Figure 2-18   Geothermal power plants. Reproduced with permission of Fotosearch.

therefore minimizing maintenance. The clean steam is converted into electrical energy much the same way as typical fossil-fueled steam turbine plants.

While geothermal is considered to be a good renewable resource of reliable power, some are concerned that the long-term effects on the use of this geothermal resource for power plants may reduce over time (i.e., dry up, reduce availability, or lose pressure). A typical geothermal power plant is shown in Figure 2-18.

Biopower

Another form of renewable energy is called "*biopower*." Biopower comes primarily from wood and agricultural residues that are burned as a fuel. Biopower has been growing steadily over the years and contributes about 12% of all renewable energy

U.S. Biopower Electricity Generation Sources

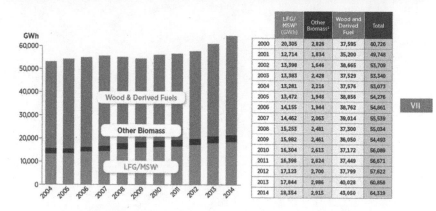

	LFG/MSW[1] (GWh)	Other Biomass[2]	Wood and Derived Fuel	Total
2000	20,305	2,826	37,595	60,726
2001	12,714	1,834	35,200	49,748
2002	13,398	1,646	38,665	53,709
2003	13,383	2,428	37,529	53,340
2004	13,281	2,216	37,576	53,073
2005	13,472	1,948	38,856	54,276
2006	14,155	1,944	38,762	54,861
2007	14,462	2,063	39,014	55,539
2008	15,253	2,481	37,300	55,034
2009	15,982	2,461	36,050	54,493
2010	16,304	2,613	37,172	56,089
2011	16,398	2,824	37,449	56,671
2012	17,123	2,700	37,799	57,622
2013	17,844	2,986	40,028	60,858
2014	18,354	2,915	43,050	64,319

Source: EIA
Reported values may vary from those included in previous versions of the Data Book due to retroactive changes by EIA.
[1] LFG = landfill gas; MSW = municipal solid waste.
[2] Other includes biogenic municipal solid waste, landfill gas, sludge waste, agricultural byproducts, and other biomass.

Biopower | November 2015 | 79

Figure 2-19   Biopower growth in the United States. Reproduced with permission of National Renewable Energy Laboratory.

generated in the United States, or about 1.6% of the total U.S. electrical generation from all sources.

Biomass is organic matter derived from living, or recently living organisms that can be used as a source of energy in combustion or indirectly in the form of a biofuel. The NREL chart shown in Figure 2-19 provides the growth picture in the United States for biopower.

CHAPTER 3

TRANSMISSION LINES

CHAPTER OBJECTIVES

After completing this chapter, the reader will be able to:

- ☑ *Explain why high-voltage transmission lines are used*
- ☑ *Explain the different conductor types, sizes, materials, and configurations*
- ☑ *Describe the electrical design characteristics of transmission lines (insulation, air gaps, lightning performance, etc.)*
- ☑ *Explain the differences between ac versus dc transmission line design, reliability, applications, and benefits*
- ☑ *Discuss overhead and underground transmission systems*

TRANSMISSION LINES

Why transmission lines? The best way to answer that question is that high-voltage transmission lines transport power long distances much more efficiently than lower-voltage distribution lines for two main reasons. First, high-voltage transmission lines take advantage of the power equation. That is, power is equal to the voltage times current. Therefore, increasing the voltage decreases the amount of current need to transport the same amount of power. Since current squared is the primarily factor in calculating power losses, lowing the current drastically reduces transportation losses. Second, raising the voltage to lower the current allows one to use smaller conductor sizes, or have more conductor capacity available for growth.

Figure 3-1 shows a three-phase 500 kV transmission line with two conductors per phase. The two conductors per phase option is called *bundling*. Power companies bundle multiple conductors together per phase in order to double, triple, or greater to increase the power transport capability of a power line, lower losses and improve other operating characteristics of the line such as electromagnetic fields and audible noise.

The type of insulation used in this line is referred to as *V-string* insulation. V-string insulation, compared to *I-string* insulation provides stability in wind

Electric Power System Basics for the Nonelectrical Professional, Second Edition. Steven W. Blume.
© 2017 by The Institute of Electrical and Electronics Engineers, Inc. Published 2017 by John Wiley & Sons, Inc.

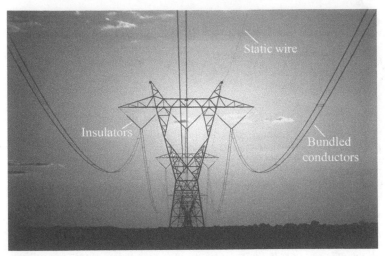

Figure 3-1   Transmission line. Reproduced with permission of Photovault.

conditions. This line also has two *static wires* on the very top to shield itself from lightning. The static wires in this case do not have insulators, instead they are directly connected to the metal towers so that lightning strikes are immediately grounded to earth. Hopefully this shielding will keep the main power conductors isolated from experiencing high-voltage transients caused by direct lightning strikes.

Raising Voltage to Reduce Current

Raising voltage to reduce current reduces conductor size and increases insulation requirements. Let us look at the power equation again:

$$Power = Voltage \times Current$$

$$Voltage_{In} \times Current_{In} = Voltage_{Out} \times Current_{Out}$$

From the power equation above, raising the voltage means the current can be reduced for the same amount of power. The purpose of step-up transformers at power plants, for example, is to increase the voltage to lower the current for power transport over long distances. Then at the receiving end of the transmission line, step-down transformers are used to reduce the voltage for easier distribution.

For example, the amount of current needed to transport 100 MW of power at 230 kV is half the amount of current needed to transport 100 MW of power at 115 kV. In other words, doubling the voltage cuts the required current in half.

High-voltage transmission lines require larger structures with longer insulator strings in order to have greater air gaps and needed insulation. However, it is usually much cheaper to build larger structures and wider right of ways for high-voltage transmission lines than it is to pay the continuous cost of high losses associated with lower-voltage power lines. Also, to transport a given amount of power from point "a" to point "b", a higher-voltage line can require much less right of way land than its equivalent multiple lower-voltage lines.

Raising Voltage to Reduce Losses

The cost for losses decreases dramatically when the current is lowered. The power losses in conductors are calculated by the formula I^2R. If the current (I) is doubled, the power losses quadruple for the same amount of conductor resistance (R)! Again, it is much more cost effective to transport large quantities of electrical power over long distances using high-voltage transmission lines because the current is less and the losses are much less. For example, when the current is cut in half, the losses are cut to one-fourth of what they would have been based on losses being a function of the current squared. In other words, doubling the current quadruples the losses. Losses are expensive to generate and utilize precious/valuable generation capacity.

Bundled Conductors

Bundling conductors significantly increases the power transfer capability of the line. The extra relatively small cost to add bundled conductors to a new line under construction is easily justified since bundling conductors significantly increases the power transfer capability of the line and reduces losses. (Note there is a practical and electrical phenomena limit to the total number of conductors in a bundle.) For example, assume that a right of way for a particular new transmission line has been secured. Designing transmission lines to have multiple conductors per phase significantly increases the power transport capability of that line for a minimal extra overall cost.

CONDUCTORS

Conductor material, type, size, and current rating characteristics are key factors in determining the power handling capability of transmission lines, distribution lines, transformers, service wires, etc. A conductor heats up when current flows through it from its resistance. The resistance per mile is constant for a conductor. The larger the diameter of the conductor the less resistance there is to current flow.

Conductors are rated by how much current causes them to heat up to a predetermined amount of degrees above ambient temperature. The amount of temperature rise above ambient (i.e., when no current flows) determines the current rating of a conductor. For example, when a conductor reaches 70°C above ambient, the conductor is said to be at full-load rating. The power company selects the temperature rise above ambient to determine acceptable conductor ratings. The power company might adopt a different current rating for emergency conditions or when the weather is extremely cold.

The amount of current that causes the temperature to rise depends on the type and size of the conductor material. The conductor type determines its strength and application in electric power systems. Let us review some of the common conductor materials used on power lines.

Conductor Material

Utility companies use different conductor materials for different applications. Copper, aluminum, and steel are the primary conductor materials used in electrical power systems. Other types of conductors such as silver and gold are actually better conductors of electricity; however, cost prohibits wide use of these materials.

Copper

Copper is an excellent conductor and is very popular. Copper is very durable and is not affected significantly by weather.

Aluminum

Aluminum is a good conductor but not as good or as durable as copper. However, aluminum costs less. Aluminum is rust resistant and weighs much less than copper.

Steel

Steel is a poor conductor when compared to copper and aluminum, however it is very strong. Steel strands are often used as the core in aluminum conductors to increase the tensile strength of the conductor.

Conductor Types

Power line conductors are either solid or stranded. Rigid conductors such as hollow aluminum tubes are used as conductors in substations because of the added strength against sag in low-profile substations when the conductor is only supported at both ends. Rigid copper bus bars are commonly used in low-voltage switch gear because of their high current rating and relatively short distances.

The most common power line conductor types are shown below:

Solid

Solid conductors, Figure 3-2, are typically smaller and stronger than stranded conductors. Solid conductors are usually more difficult to bend and are easily damaged.

Stranded

As shown in Figure 3-3, *stranded* conductors have three or more strands of conductor material twisted together to form a single conductor. Stranded conductors can carry high currents and are usually more flexible than solid conductors.

Figure 3-2   Solid conductor.

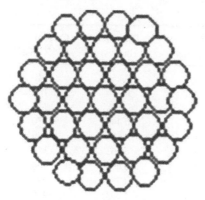

Figure 3-3   Stranded conductor.

Aluminum-Conductor Steel-Reinforced (ACSR)

To add strength to aluminum conductors, Figure 3-4 shows steel strands that are used as the core of aluminum stranded conductors. These high-strength conductors are normally used on long span distances, for minimum sag applications.

Conductor Size

There are two conductor size standards used in electrical systems. One is for smaller size conductors (*American Wire Gauge*) and the other circular mils is for larger conductor sizes. Table 3-1 compares common conductor sizes and standards.

American Standard Wire Gauge (AWG)

The American Standard Wire Gauge is an old standard that is used for relatively smaller conductor sizes. The scale is in reverse order; in other words, the numbers get smaller as the conductors get larger.

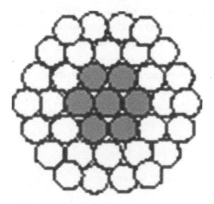

Figure 3-4   ACSR conductor.

TABLE 3-1 Typical ACSR Conductor Sizes

Cross section (inches)	Size (AWG or cmils)	Size Copper Equivalent	Ratio (Al to Steel)	Diameter (inches)	Current (A) (75°C rise)
0.250	4	6	7/1	0.250	140
0.325	2	4	6/1	0.316	180
0.398	1/0	2	6/1	0.398	230
0.447	2/0	1	6/1	0.447	270
0.502	3/0	1/0	6/1	0.502	300
0.563	4/0	2/0	6/1	0.563	340
0.642	266,000	3/0	18/1	0.609	460
0.783	397,000	250,000	26/7	0.783	590
1.092	795,000	500,000	26/7	1.093	900
1.345	1,272,000	800,000	54/19	1.382	1200

Circular Mils

The circular mils standard of measurement is used for large conductor sizes. Conductors greater than AWG 4/0 are measured in circular mils. One circular mil is equal to the area of a circle having a 0.001 inch (1 mil) diameter. For example, the magnified conductor in Figure 3-5 has 55 circular mils. In actual size, a conductor of 55 circular mils is about four times smaller than the period at the end of this sentence. Therefore, conductors sized in circular mils are usually stated in the thousands of circular mils (i.e., kcm).

Table 3-1 shows typical conductor sizes and associated current ratings for outdoor bare ACSR conductors having a current rating of 75°C rise above ambient. The table also shows the equivalent copper size conductor.

Insulation and Outer Covers

These metal wire current carrying conductors can be insulated or non-insulated when in use. Non-insulated conductors (i.e., bare wires) normally use what is called "*insulators*" as the means for separating the bare wires from the grounded structures, making air their insulation. Insulated conductors use plastic, rubber, or other jacketing materials for electrical isolation. High-voltage insulated conductors are normally used in underground systems. Insulated low-voltage service wires are often used for residential overheard and underground lines.

In the 1800s, Ronalds, Cooke, Wheatstone, Morse, and Edison made the first insulated cables. The insulation materials available at that time were natural

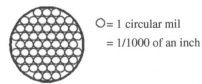

Figure 3-5   Circular mils.

TABLE 3-2   Transmission Voltages

Voltage Class	Voltage Category	System Voltage
69,000		Sub-transmission
115,000		
138,000		
161,000		Transmission
230,000	Extra high voltage (EHV)	
345,000		
500,000		
765,000		
Above 1,000,000	Ultra high voltage (UHV)	

substances such as cotton, jute, burlap, wood, and oil-impregnated paper. With the development of rubber compounds and the invention of plastic, insulation for underground cables have become much more reliable, cost effective, and efficient.

Voltage Classes

Table 3-2 shows the various transmission and sub-transmission system voltages used in North America. This table is not absolute; some power companies designate their system voltages a little differently. Note it is quite common to use sub-transmission voltages to transport power medium distances (i.e., across large populated areas) or to transport power long distances if the total current requirement is low, such as serving less populated areas that are far away.

The higher transmission system voltages tend to be more standardized compared to the lower distribution voltages. There are many subtle variations in distribution voltages than transmission.

Voltage class is the term often used by equipment manufacturers and power companies to identify the voltage that the equipment will be connected to. A manufacturer might use the voltage class to identify the intended system operating voltage for their equipment. A power company might use the voltage class as a reference to the system discussed in a conversation. A voltage class might include several *nominal* operating voltages. Nominal voltages are the everyday normal actual voltages. For example, a circuit breaker might be a 125 kV voltage class piece of equipment that is operating at a nominal 115 kV voltage.

Voltage category is often used to identify a group of voltage classes. For example, "extra high voltage" (or EHV) is a term used to state whether an equipment manufacturer builds transmission equipment versus distribution equipment which would be categorized as "high voltage equipment" (or HV).

System voltage is a term used to identify whether distribution, transmission, or secondary is referenced. For example, power companies normally distinguish between distribution and transmission departments. A typical power company might distinguish between distribution line crews, transmission line crews, etc. Secondary system voltage usually refers to customer service voltages.

TRANSMISSION LINE DESIGN PARAMETERS (OPTIONAL SUPPLEMENTARY READING)

This section discusses in more detail the design parameters regarding high-voltage transmission lines.

Insulation

The minimum insulation requirements for a transmission line are determined by first evaluating individually, the minimum requirements for each of the following factors:

Any of the insulation criteria listed below could dictate the minimum spacing and insulation requirements for the transmission line.

Air Gaps for 60 Hz Power Frequency Voltage

Open air has a flashover voltage rating. A rule of thumb is one foot of air gap for every 100 kV of voltage. Detailed reference charts are available to determine the proper air-gap requirements based on operating voltage, elevation, and exposure conditions.

Contamination Levels

Transmission lines located near oceans, alkali salt flats, cement factories, etc. require extra insulation for lines to perform properly under contamination-prone environments. Salt mixing with moisture, for example, can cause leakage currents and possible undesirable insulation flashovers to occur. Extra insulation is often required for these contamination prone environments. This extra insulation could increase the minimum air-gap clearance.

Expected Switching Surge Overvoltage Conditions

When power system circuit breakers operate, or large motors start, or disturbances happen on the power grid, transient voltages could occur that can flashover the insulation or air gap. The design engineer studies all possible switching transient conditions to make sure adequate insulation is provided on the line at all times.

Safe Working Space

The National Electrical Safety Code (NESC) specifies the minimum phase-to-ground and phase-to-phase air-gap clearances for all power lines and substation equipment. These NESC clearances are based on safe working space requirements. In some cases, the minimum electrical air-gap clearance is increased to meet NESC requirements.

Lightning Performance

Transmission lines frequently use shield wires to improve the line's operating performance under lightning conditions. These *shield wires* (sometimes called *static wires* or *earth wires*) serve as a high-elevation ground wires to attract lightning. When lightning strikes the shield wire, surge current flows through the wires, towers, ground rods, and finally into the earth where the energy is dissipated. Sometimes extra air-gap clearance is needed in towers to overcome the possibility of the tower flashing back over to the power conductors when lightning energy is being dissipated. This condition is mitigated by good tower grounding practices.

Audible Noise

Audible noise can also play a role in designing high-voltage power lines. Audible noise from foul-weather, generated by electrical stress, corona discharge, and the low-frequency hum can become troublesome if not evaluated during the design process. There are ways to minimize audio noise, most of which tend to increase conductor size and/or air-gap spacing.

UNDERGROUND TRANSMISSION
(OPTIONAL SUPPLEMENTARY READING)

Underground transmission is usually 3–10 times costlier than overhead due to right of way, obstacles, and material costs. It is normally used in urban areas or near airports where overhead is not an option. Most modern underground transmission cables are made of solid dielectric polyethylene materials and can have ratings in the order of 400 kV. Figure 3-6 shows a 230 kV underground transmission line.

Figure 3-6   Underground transmission.

Figure 3-7   Overhead transmission.

DC TRANSMISSION SYSTEMS
(OPTIONAL SUPPLEMENTARY READING)

DC transmission systems are used for economic reasons, system synchronization benefits, interconnecting 60 Hz with 50 Hz systems, and power flow control. The three-phase ac transmission line is converted into a 2-pole (plus and minus) dc transmission line using bidirectional rectification *converter stations* at both ends of the dc line. The converter stations convert the ac power into dc power and vice versa. The reconstructed ac power must be filtered for improved power quality performance before connected to the ac system.

DC transmission lines do not have phases, instead they have positive and negative *poles*. The Pacific Northwest dc transmission line shown in Figure 3-7, for example, operates at ± 500 kV or 1 million Volts pole-to-pole. There are no synchronization issues with dc lines. The frequency of dc transmission is zero and therefore no concerns for variations in frequency between interconnected systems. A 60 Hz system can be connected to a 50 Hz system using a dc line.

For economic reasons, the dc line may have advantages over the ac line in that the dc lines have only two conductors versus three conductors in ac lines. The overall cost to build and operate a dc line including converter stations may cost less than an equivalent ac line due to the savings in one less conductor, narrower right of ways, and less expensive towers. This is usually the situation for lines longer than 300 miles in length.

CHAPTER 4

SUBSTATIONS

CHAPTER OBJECTIVES

After completing this chapter, the reader will be able to:

- ☑ *Identify and describe the operations of all major equipment used in substations*
- ☑ *Explain the operation and need for voltage regulators and tap changers*
- ☑ *Discuss how circuit breakers and disconnect switches are used to isolate equipment*
- ☑ *Explain the purpose of capacitors, reactors, and static VAR compensators used in electric power systems*
- ☑ *Describe effective preventive maintenance programs used for substation equipment*

SUBSTATION EQUIPMENT

The major types of equipment found in most transmission and distribution substations are discussed in this chapter. The purpose, function, design characteristics and key properties are all explained. After the equipment is discussed and planned, essential predictive maintenance techniques are discussed. The reader should get a good fundamental understanding of all the important aspects of major equipment found in substations and how they are used and operated.

The substation equipment discussed in this chapter include:

- Transformers
- Regulators
- Circuit Breakers and Reclosers
- Air Disconnect Switches
- Lightning Arresters
- Electrical Bus
- Capacitor Banks

Electric Power System Basics for the Nonelectrical Professional, Second Edition. Steven W. Blume.
© 2017 by The Institute of Electrical and Electronics Engineers, Inc. Published 2017 by John Wiley & Sons, Inc.

- Reactors
- Static VAR Compensators
- Control Building

TRANSFORMERS

Transformers are essential components in electric power systems. They come in all shapes and sizes. Power transformers are used to convert high-voltage power to low-voltage power and vice versa. Power can flow in both directions in a transformer; from the high-voltage side to the low-voltage side or from the low-voltage side to the high-voltage side. Generation plants use large *step-up transformers* to raise the voltage of the generated power for efficient transport of power over long distances. Then *step-down transformers* convert the power to sub-transmission voltage levels as shown in Figure 4-1 or distribution voltages as in Figure 4-2 for further transport power for consumption. *Distribution transformers* are mounted on distribution overhead poles or underground padmounts to further convert distribution voltages down to suitable voltages for residential, commercial, and industrial consumption (see Figure 4-3).

There are many types of transformers used in electric power systems. Through the use of scale factors, *Instrument transformers* are used to connect high-power equipment to low-power electronic instruments for monitoring system voltages, currents, and power at convenient levels. Instrument transformers include current transformers (*CTs*) and potential transformers (PTs)). These instrument transformers connect to metering, protective relaying, and telecommunications equipment. *Regulating transformers* are used to maintain proper distribution voltages so that consumers have

Figure 4-1   Step-down transformer.

Figure 4-2   Distribution power transformer.

stable wall outlet voltage. *Phase shifting transformers* are used to control power flow over interconnection tie lines.

Transformers can be single phase, three phase, or *banked* together to operate as a single unit. Figure 4-3 shows a three-phase transformer bank.

Figure 4-3   Transformer bank.

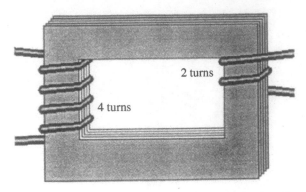

Figure 4-4   Transformer windings. Courtesy of Alliant Energy.

Transformer Fundamentals

Transformers work by combining the two physical laws that were discussed earlier in Chapter 1. As a review; physical law #1 states that a voltage is produced on any conductor in a changing magnetic field. Physical law #2 states that a current flowing in a wire produces a magnetic field. Transformers combine these two principles by using two coils of wire, having a changing voltage source and current corresponding to load. The current flowing in the coil on one side of the transformer induces a voltage in the coil on the other side. (Hence, the two coils are coupled together by magnetic fields.)

This is a very important concept because the entire electric power system depends on these relationships. Looking at them closely; the voltage on the opposite side of a transformer is proportional to the *turns ratio* of the transformer. And, the current on the other side of the transformer is inversely proportional to the turns ratio of the transformer.

For example, the transformer in Figure 4-4 has a turns ratio of 2:1.

Suppose the 2:1 turns ratio transformer in Figure 4-4 has 240 Vac at 1 A applied on its primary winding (left side); it will produce 120 Vac at 2 A on its secondary winding (right side) as seen in Figure 4-5. Note, power equals 240 W on either side (i.e., voltage × current). As discussed earlier, raising the voltage (i.e., like on transmission lines) lowers the current and thus significantly lowers system losses.

Power Transformers

Figure 4-6 shows the inside of a large power transformer. Power transformers consist of two or more windings for each phase and these windings are usually wound around

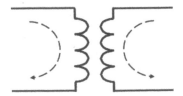

Figure 4-5   Transformer turns ratio.

Figure 4-6   Transformer core and coils. Courtesy of ABB.

an *iron core*. The iron core improves the efficiency of the transformer by concentrating the magnetic field and reducing transformer losses. The high-voltage and low-voltage windings have a unique number of coil turns. The turns ratio between the coils dictates the voltage and current relationships between the high- and low-voltage sides.

Bushings

Bushings are used on transformers, circuit breakers, and many other types of electric power equipment as connection points. Bushings connect outside conductors to conductors inside the equipment. Bushings provide insulation between the energized conductor and the grounded metal tank surrounding the conductor. The conductors inside the bushings are normally solid copper rods surrounded by porcelain insulation. Usually an insulation dielectric such as oil or gas is added inside the bushing between the copper conductor and the porcelain housing to improve its insulation characteristics. Mineral oil and sulfur hexafluoride (SF_6) gas are common dielectric materials to increase insulation.

Note, transformers have large bushings on the high-voltage side of the unit and small bushings on the low-voltage side. In comparison, circuit breakers have the same size bushings on both sides of the unit.

Figures 4-7 and 4-8 are examples of typical transformer bushings. Notice the oil level through the glass portion at the top of the bushing. Sometimes oil level gauges are used for oil level inspections.

The part of the bushing that is exposed to the outside atmosphere generally has *skirts*, or sometimes called *ribs* to reduce unwanted leakage currents. The purpose of skirts is to increase the leakage current distance in order to decrease the leakage current. Cleanliness of the outside porcelain is also important. Contaminated or dirty bushings can cause arcing that can result in flashovers especially during light rain or fog conditions.

Figure 4-7   Bushing oil level gauge.

Figure 4-8   Transformer bushing.

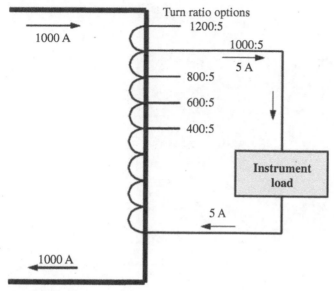

Figure 4-9   CT connections.

Instrument Transformers

The term *instrument transformer* refers to current and voltage transformers that are used to scale down actual power system quantities for metering, protective relaying, and/or system monitoring equipment. The application of both current and potential transformers provides scaled-down quantities for power and energy information.

Current Transformers

Current transformers or *CTs* are used to scale down the high magnitude of current flowing in the main high-voltage conductors to a level much easier to work with safely. For example, it is much easier to work with 5 A of current in the CTs secondary circuit than it is to work with 1000 A of current in the CTs primary circuit.

Figure 4-9 shows a typical CT connection diagram. Using the CTs turn ratio as a *scale factor* provides the current level required of the monitoring instrument; yet the current located in the high-voltage conductor is actually being measured.

Taps (or connection points to the coil) are used to allow options for various turns ratio scale factors to best match the operating current with the instrument's current requirements.

Most CTs are located on transformer and circuit breaker bushings as shown in Figure 4-10. Figure 4-11 shows a low-voltage CT. Figure 4-12 shows a standalone high-voltage CT.

Potential Transformers

Similarly, *potential transformers* (*PTs*) are used to scale down very high voltages to levels that can be worked safely. For example, it is much easier to work with 115 Vac

Figure 4-10   Bushing CT.

than 69 kVac. Figure 4-13 shows how a 69 kV PT is connected. The 600:1 turns ratio or scale factor is taken into account in the calculations of actual voltage. PTs are also used for metering, protective relaying, and system monitoring equipment. The instruments connected to the secondary side of the PT are programmed to account for the turns ratio scale factor.

Like most transformers, taps are used to allow options for various turns ratios to best match the operating voltage with the instrument's voltage level requirements.

Figure 4-11   Low voltage CT.

Figure 4-12   External HV CT.

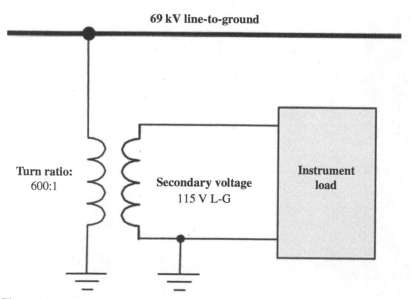

69 kV line-to-ground

Turn ratio:
600:1

Secondary voltage
115 V L-G

**Instrument
load**

Figure 4-13   PT connections.

Figure 4-14   Low-voltage PT. Courtesy of Alliant Energy.

An example of a low-voltage PT is shown in Figure 4-14 and a high-voltage PT in Figure 4-15.

Autotransformers (Optional Supplementary Reading)

Autotransformers are a special construction variation to regular two winding transformers. Autotransformers share a winding. Single-phase, autotransformers

Figure 4-15   High-voltage PT. Courtesy of Alliant Energy.

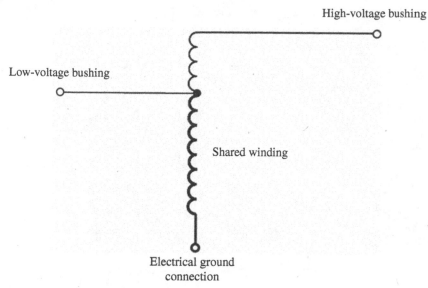

Figure 4-16   Autotransformer.

contain a primary winding and a secondary winding on a common core. However, part of the high-voltage winding is shared with the low-voltage winding, as shown in Figure 4-16.

Autotransformers work best with small turns ratios (i.e., less than 5:1). Autotransformers are normally used for very high voltage transmission applications. For example, autotransformers are commonly found matching 500–230 kV or 345–120 kV system voltages. Material cost savings is an advantage of autotransformers. Size reduction is another advantage of autotransformers.

Figure 4-16 shows how an autotransformer is connected. The physical appearance looks the same as any other power transformer. A person needs to review the transformer nameplate to tell whether it is an autotransformer or a conventional transformer.

Note, under "no-load" conditions, high-side voltage will be the sum of the primary and shared winding voltages, and the low-side voltage will be equal to the shared winding voltage.

REGULATORS

It is important for electric utility companies to provide their customers with regulated or steady voltage at all times, otherwise several undesirable conditions might occur. Normally, residential 120 Vac is *regulated* to ±5% (i.e., 126 Vac ↔ 114 Vac). The first residential customer outside the substation should not have voltage exceeding 126 Vac and the last customer at the end of the distribution feeder should not have

Figure 4-17   Three-phase regulator.

voltage less than 114 Vac. Power companies try to regulate the distribution voltage to be within a nominal 124–116 Vac.

Customer service problems can occur if voltages are too high or too low. For example, low voltage can cause motors to overheat and burn out. High voltages can cause light bulbs to burn out too often or cause other appliance issues. Utility companies use voltage regulators to keep the voltage level within an acceptable or controlled range or bandwidth.

Voltage regulators are similar to transformers. Regulators have several taps on their windings that are changed automatically under load conditions by a motor-driven control system called the *load tap changer* or *LTC*. Figure 4-17 shows a substation three-phase voltage regulator and Figure 4-18 shows a single-phase regulator. Three single-phase regulators can be used in a substation or out on a distribution line.

Theory of Operation

Normally a regulator is specified as being ±10%. The distribution feeder voltage out of the substation regulator can be raised up to 10% or lowered down to 10%. There are 16 different tap positions on either the raise or lower sides of the neutral position. There is a reversing switch inside the LTC that controls whether to use the tap connections to raise the output feeder voltage or use them to lower the output feeder voltage. Therefore, the typical voltage regulator has "*33 positions*" (i.e., 16 raise, 16 lower plus neutral). Figure 4-19 shows the 33 positions on the dial. Each position can change the primary distribution voltage by 5/8% (i.e., 10% divided by 16 taps).

For example: A typical 7200 V, ±10% distribution regulator would have 33 tap positions. Each tap could raise or lower the primary distribution voltage 45 V (i.e., 10% of 7200 equals 720 V, and 720 V divided by 16 taps equals 45 V per tap). This

Figure 4-18   Single-phase regulator.

variation in feeder voltage gives plenty of room for customer service transformer to regulate their required 5%.

Reactor coils are used to reduce the number of actual winding taps to 8 instead of 16. Reactor coils allow the regulator's output contactor to be positioned between two winding taps at the same time for half the tap voltage. Figure 4-20 shows the tap changer mechanism with the reactor coil bridging both the first with neutral, thus providing a half tap voltage.

Line regulators are sometimes used near the end of long distribution feeders to re-regulate the voltage to the customers downstream of the substation regulator. Line regulators make it possible to extend the length of distribution feeders needed to serve customers at long distances from the substation.

Figure 4-21 shows a three-phase LTC mechanism inside a regulator. Figure 4-22 shows the switch contacts.

Figure 4-19   Regulator dial. Courtesy of Alliant Energy.

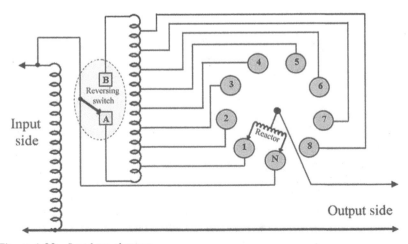

Figure 4-20   Load tap changer.

Figure 4-21   Tap changer.

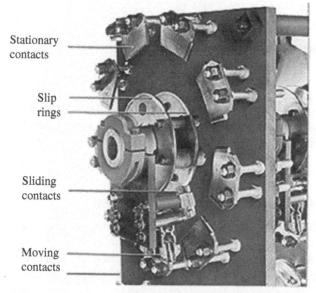

Stationary
contacts

Slip
rings

Sliding
contacts

Moving
contacts

Figure 4-22   Switch contacts. Courtesy of Alliant Energy.

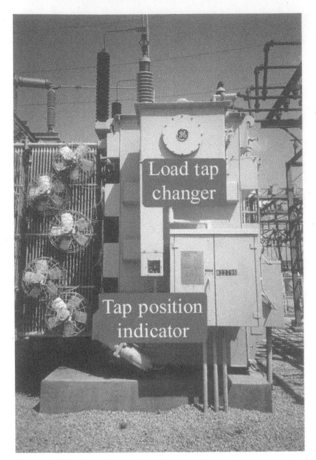

Figure 4-23   Load tap changing transformer.

Figure 4-23 shows a *load tap changing transformer* (LTC transformer). LTC transformers combine a step-down transformer with a voltage regulator. LTC transformers offer cost saving advantages. However, two LTC transformers are normally required per substation in order to have load transfer capabilities for regulator maintenance purposes.

Regulator Controls (*Optional Supplementary Reading*)

Voltage regulators use an electronic control scheme to automatically operate the raise/lower tap changer. A potential transformer (PT) is used to input actual voltage to the control circuits. A current transformer is used to determine the amount of load on the regulator. The control circuit constantly monitors the voltage level on the regulated side and sends commands to the motor operator circuit of the tap changer to a raise or lower the regulated voltage based on the control settings. The control settings are programmable by the engineer. The common settings are as follows:

Base Voltage

This is the desired voltage reference setting used to establish the regulator's *base output voltage* (e.g., 122 V is common). When the regulator PT senses the output voltage to be above or below this base setting, the tap changer motor is commanded to raise or lower the output voltage until it comes into the desired bandwidth of the base voltage.

Bandwidth

The base voltage *bandwidth* setting controls the amount of voltage tolerance above and below the base voltage setting. The regulator does not change taps unless the actual output voltage goes outside this bandwidth setting (e.g., 2 V bandwidth is normal). For example, if the base voltage is set for 122 Vac, the distribution voltage would have to rise above 124 Vac to cause a command to lower the regulated voltage. Similarly, the distribution regulator voltage would have to go below 120 Vac to cause the LTC to raise the regulated voltage.

Time Delay

The *time delay* setting prevents momentary voltage changes and therefore reduces the wear and tear of the LTC. For example, the actual distribution voltage would have to exceed the bandwidth for the duration of a preset time delay (i.e., 60 seconds) before the motorized tap changer would begin to operate.

Manual/Auto

For safety purposes, the *manual/auto switch* is used to disable the automatic control of the regulator for personnel working on associated equipment.

Compensation

The *compensation* setting is used to control voltage regulation based on conditions some distance down the line. Suppose the mail load area to be regulated is 10 miles from the regulator. The compensation settings are adjusted so that voltage source is regulated based on the conditions occurring 10 miles away. This can lead to high voltage problems for the first customer outside of the substation. The control is set to compensate for an estimated voltage drop on the distribution line.

CIRCUIT BREAKERS

The purpose of a *circuit breaker* is to interrupt current flowing in a line, transformer, bus, or other equipment when a problem occurs and the power has to be turned off. Current interruption can be normal, load current being too high, high *fault current* (short circuit current or problem in the system) or simply tripped by protective relaying equipment in anticipation of an undesirable event or disturbance. A breaker accomplishes this by mechanically moving electrical contacts apart inside an *interrupter*, causing an arc to occur that is immediately suppressed by the high dielectric medium inside the interrupter. Circuit breakers are triggered to open or close by protective relaying equipment using the substation battery system.

The most common types of *dielectric* medium used to extinguish the arc inside the breaker interrupter are listed below:

- *oil* (clean mineral)
- *gas* (SF_6 or sulfur-hexafluoride)
- *vacuum*
- *air*

These dielectric mediums also classify the breaker such as oil circuit breaker (OCB), gas circuit breaker (GCB), and power circuit breaker (PCB).

Other comments regarding circuit breakers include: compared to fuses, circuit breakers have the ability to open and close repeatedly while a fuse opens the circuit one time and must be replaced. Fuses are single-phase devices, whereas breakers are normally gang operated three-phase devices. Breakers can interrupt very high magnitudes of current. Breakers can close into a fault and trip open again. Breakers can be controlled remotely. Breakers need periodic maintenance.

Oil Circuit Breakers

The *oil circuit breaker* (sometimes called *OCB*) interrupts arcs in clean mineral oil. The oil provides a high resistance between the opened contacts to stop current flow. Figure 4-24 shows an OCB. The interrupting contacts (referred to as *interrupter*) are

Figure 4-24   Oil circuit breaker.

Figure 4-25   Interrupter contacts.

inside the oil filled tanks. Inspection plates are provided to allow close view of the interrupter contacts to determine maintenance requirements.

Oil circuit breakers have the ability to be used in systems that range from low to very high voltage. Oil has a high dielectric strength compared to air. Bushings are usually angled to allow large conductor clearances in the open air areas and smaller clearances in the oil encased areas. The main disadvantage of using oil is the environmental hazard if spilled. A maintenance concern for oil breakers is that the oil becomes contaminated with gases during arc suppression. The oil must be filtered or replaced periodically or after a specified number of operations to insure the oil retains a high dielectric strength.

Figure 4-25 shows a single tank three-phase oil breaker's interrupter contacts. Note the wide conductor spacing for the air components and the small conductor spacing in the oil-immersed components. The operating voltage of this breaker is low enough to have all three phases in one tank.

SF₆ Gas Circuit Breakers

Sulfur hexafluoride gas breakers (sometimes called SF_6 or *GCB*s) have their contacts enclosed in a sealed interrupting chamber filled with SF_6 gas. SF_6 gas is a nonflammable inert gas which has a very high dielectric strength, much greater than oil. Inert gases are colorless, odorless, and tasteless and are difficult to form other chemical compounds. These properties enable the breaker to interrupt current quickly and

Figure 4-26   Gas circuit breaker.

maintain relatively small equipment dimensions. The operating disadvantage of using SF_6 gas circuit breakers is that the gas turns to liquid at $-40°C$ or $-40°F$. Maintaining correct gas pressure is also an operational concern. Heaters are usually wrapped around the interrupter chambers in extreme cold weather environments to maintain proper temperature and pressure. Figures 4-26, 4-27, and 4-28 are photos of SF_6 gas circuit breakers

Vacuum Circuit Breakers

Vacuum circuit breakers (i.e., *VCBs*) extinguish the arc by opening the contacts in a vacuum. (Vacuum has a lower dielectric strength than oil or gas, but higher than air.) These circuit breakers are smaller and lighter than air circuit breakers and typically are found in distribution voltage substations or switchgear under 35 kV. Figure 4-29 shows a typical VCB.

The contacts are enclosed in an evacuated bottle where no rated current can flow when the contacts separate. When the breaker opens, the arc is put out simply and quickly.

Air Circuit Breakers

Since the dielectric strength of air is much less than oil or SF_6 gas, *air breakers* are relatively large and are usually found in lower-voltage installations such as "metal-clad" switchgear. Figure 4-30 shows a 12 kV air breaker used in switch gear.

Figure 4-27   345 kV gas breaker.

Figure 4-28   161 kV gas breaker. Courtesy of Alliant Energy.

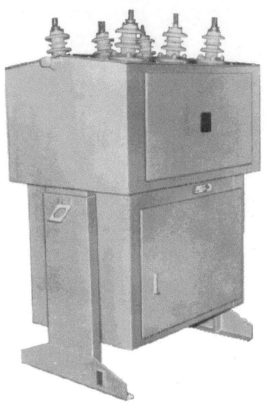

Figure 4-29   Vacuum circuit breaker. Courtesy of Alliant Energy.

The very high voltage *air-blast* circuit breaker (not shown) is another type of circuit breaker that is used for sub-transmission voltages. Air-blast breakers use a compressed blast of air across the interrupting contacts to help extinguish the arc. Most air-blast circuit breakers are considered old or obsolete and have been replaced.

RECLOSERS

Similar to breakers, *reclosers* provide circuit breaker functionality and they also include basic system protective relaying equipment to control the automatic opening and reclosing of distribution feeder circuits. Reclosers are most commonly used on distribution systems. They offer cost advantages over standard circuit breakers that require separate protective relaying equipment.

The recloser's incorporated protective relaying equipment can be programmed to trip at specific overcurrent conditions and reclose at specific time intervals. After a circuit trip and a programmable time delay, the recloser automatically re-energizes the circuit. (Please be advised that the automatic reclosing feature on reclosers can be deactivated while crew members are performing maintenance on the line.) Also,

Figure 4-30   Air circuit breaker. Courtesy of Alliant Energy.

reclosers are responsible for re-energizing a feeder after lightning strikes, car-pole accidents, etc.

Reclosers are commonly used as circuit breakers on distribution lines (see Figures 4-31 and 4-32) or in smaller substations (see Figure 4-33) having low fault currents. Reclosers are typically set to trip and reclose two or three times before a *lock-out* condition occurs. Lock-out means the recloser went through its cycle of trips and closes and locks itself open. A line working person must manually reset the recloser lock-out for power to be restored. If the fault condition clears before the recloser locks-out, the protective relaying resets back to the start of the sequence after a programmed time delay.

Reclosers can also be tripped manually or via remote control. Remote control capability is used at control centers for system operators. Newer reclosers work in conjunction with distribution automation, which will be discussed later. Also, reclosers can be used as a load break switch or sectionalizer.

DISCONNECT SWITCHES

There are many purposes for *disconnect switches* in substations and on power lines. They are used to isolate or de-energize equipment for maintenance purposes, transfer

Figure 4-31   Modern recloser.

load from one source to another in planned or emergency conditions, provide visual openings for maintenance personnel (an OSHA (Occupational Safety and Health Administration) requirement for safety against accidental energization), and other reasons. Disconnect switches have low current interrupting ratings compared to circuit breakers. Normally power lines are first de-energized by circuit breakers (due to their high-current interrupting capability) followed by the opening of the air disconnect switches for isolation.

Substations

There are many types of substation disconnect switches such as *vertical break* and *horizontal break* types. Disconnect switches are normally *gang* operated. The term gang is used when all three phases are operated with one operating device. Air disconnect switches are usually opened and closed using control handles mounted at the base of the structure. Sometimes motor operator mechanisms are attached to the control rods to remotely control their operation. A vertical break switch is shown in Figure 4-34 and a horizontal break switch is shown in Figure 4-35.

Figure 4-32   Older distribution line recloser. Courtesy of Alliant Energy.

Some disconnect switches such as the one shown in Figure 4-36 use spring loaded devices called *arcing rods* to help clear arcs from small currents by whipping open the electrical connection after the switch's main contacts have opened. This spring loaded device is also referred to as a *whip* or *arcing horn*. The arcing rods increase the switch's current opening rating, but usually not enough to open normal load. They might open a long unloaded line or perhaps a paralleling load transfer operation. Also, arching rods are sacrificial, in that the rods get pitted in the opening process, rather than the main switch contacts. Rods are cheap and easy to replace.

Line Switches

Line disconnect switches are normally used to isolate sections of line or transfer load from one circuit to another. The picture in Figure 4-37 is an example of a typical sub-transmission line switch. This particular switch incorporates *vacuum bottles* to help extinguish arcs from interrupting light load currents.

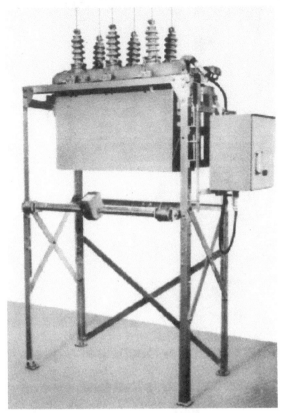

Figure 4-33   Older substation recloser. Courtesy of Alliant Energy.

Figure 4-34   Vertical break. Courtesy of Alliant Energy.

Figure 4-35   Horizontal break. Courtesy of Alliant Energy.

LIGHTNING ARRESTERS

Lightning arresters are designed to limit the line-to-ground voltage in the event of lightning or the occurrence of other excessive transient voltage conditions. Some of

Figure 4-36   Arcing rods.

Figure 4-37   Line switch.

the older gap type lightning arresters actually short circuit the line or equipment causing the circuit breaker to trip. The breaker would then reclose the circuit when the transient over-voltage condition goes away. The lightning arrester limits the voltage on the equipment near the lightning arrester from experiencing high-voltage transient conditions. Therefore, the insulation properties coordinate with the lightning arresters discharge voltage.

For example, suppose an 11 kV lightning arrester was installed on a 7.2 kV line-to-neutral system. The lightning arrester will conduct if the line-to-neutral voltage exceeded approximately 11 kV. Equipment connected to this distribution system might have a flashover rating of 90 kV. Therefore, the arrester clamped or limited the high-voltage transient to under the equipment's flashover rating and prevented the equipment from experiencing a flashover or insulation failure.

The newer lightning arresters use gapless *metal oxide* semi-conductor materials to clamp or limit the voltage. These newer designs offer better voltage control and have higher energy dissipation characteristics.

Aside from the voltage rating for which the arrester is applied, arresters fall into different energy dissipation classes. An arrester might have to dissipate energy up until the circuit breaker clears the line. *Station class* arresters (see Figure 4-38) are

Figure 4-38   Station class. Courtesy of Alliant Energy.

the largest types and can dissipate the greatest amount of energy. They are usually located adjacent to large substation power transformers. *Distribution class* arresters (see Figure 4-39) are generously distributed throughout the distribution system in areas known to have high lightning activity. They can be found near distribution transformers, overhead to underground transition structures, and along long distribution lines. *Intermediate class* arresters are normally used in substations that do not have excessive short circuit current. Residential and small commercial customers may use *secondary class* arresters to protect large motors, sensitive electronic equipment, and other voltage surge sensitive devices connected to their service panel.

ELECTRICAL BUS

The purpose of *electrical bus* in substations is to connect equipment together. Bus is a conductor, or group of conductors, that serve as a common connection between two or more circuits. The bus is supported by station post insulators. These insulators mount to the bus structures. The bus can be constructed of 3–6 inch rigid aluminum tubing (called "*rigid bus*") or wires with insulators on both ends, called "*strain bus.*"

The *buswork* consists of structural steel that supports the insulators that support the energized conductors. The buswork might also include the air disconnect switches. Special bus configurations allow for transferring load from one feeder to another and to bypass equipment for maintenance.

Figure 4-40 is an example of typical low-profile rigid-bus found in distribution substations.

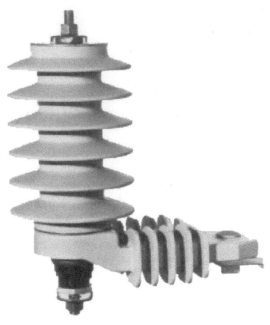

Figure 4-39   Distribution class. Courtesy of Alliant Energy.

Figure 4-40   Example of typical electrical bus.

CAPACITOR BANKS

Capacitors are used to improve the operating efficiency of electric power systems and help transmission and distribution system voltage stability during disturbances and high-load conditions. Capacitors are used to cancel out the lagging current effects from motors and transformers. Capacitors can reduce system losses and help provide voltage support. Another benefit of capacitor is how they can reduce the total current flowing through a wire thus leaving capacity in the conductors for additional load.

Capacitor banks can be left online continuously to meet the steady state reactive power requirements of the system or they can be switched on or off to meet dynamic reactive requirements. Some of the capacitor banks are switched seasonally (i.e., to accommodate air conditioning load in the summer) and others are switched daily to accommodate industrial loads.

Capacitor banks can be switched manually, automatically, locally, and remotely. For example, system control center operators commonly switch substation capacitor banks on and off to meet load requirements or system stability reactive demand requirements. Providing capacitive support maintains good system voltage, reduces system losses, and improves contingency reliability.

Substation Capacitor Banks

Figure 4-41 shows a typical substation capacitor bank. Actually, this picture shows two three-phase capacitor banks (one in the foreground and one in the background). The vertical circuit breakers on the far right of the picture provide the switching functionality of these substation capacitor banks.

Figure 4-41   Station capacitor bank.

Figure 4-42   Distribution capacitor bank.

Distribution Capacitor Bank

Capacitor banks are installed on distribution lines to reduce losses, improve voltage support, and provide additional load capacity on the distribution system (See Figure 4-42). Note, reducing distribution system losses with capacitors is very effective since that also reduces transmission losses.

The closer a capacitor is installed to the actual inductive load itself, the more beneficial. For example, if capacitors are installed right at the motor terminals at an industrial load, losses are reduced in the lines feeding the motor, distribution losses would be reduced, transmission losses would be reduced, and generation losses would also be reduced.

REACTORS

Reactor is another name for a high-voltage inductor. They are essentially one winding transformers. Reactors are used in electric power systems for two main reasons. First, reactors are used in a shunt configuration (i.e., line-to-ground connections), to help regulate transmission system voltage under light load conditions by absorbing surplus reactive power (VARs) from *line charging*. Line charging is the term used to describe the capacitance effects of long transmission lines since they are essentially long skinny capacitors (i.e., two conductors separated by a dielectric being the air). Second, they are connected in series to reduce fault current in distribution lines.

Reactors can be open air coils or coils submerged in oil. Reactors are available in either single-phase or three-phase units.

Shunt Reactors–Transmission

The electrical characteristics and performance of long high-voltage transmission lines can be improved through the use of shunt reactors. *Shunt reactors* are used on transmission lines to help lower voltage or balance reactive power flowing on the system. They can be used to absorb excess reactive power. Reactors are normally disconnected during heavy load conditions and are connected during periods of low load. Reactors are switched online during light load conditions (i.e., late at night or early morning) when the transmission line voltage tends to creep upward. Conversely, shunt capacitors are added to transmission lines during high-load conditions to raise the system voltage. In both cases, distribution regulators provide the fine adjustments to service voltage to the customers.

Another application of shunt reactors is to help lower transmission line voltage when energizing a long transmission line. For example, suppose a 200 mile, 345 kV transmission line is to be energized; the line charging effect of this long transmission line can cause the far-end voltage to be in the order of 385 kV. Switching on a shunt reactor at the far end of the line before energization can reduce the far-end voltage to approximately 355 kV. This reduced far-end voltage will result in a lower transient voltage condition when the far-end circuit breaker is closed connecting the transmission line to the system when allowing current to flow. Once load is flowing in the line, the shunt reactor can be disconnected and the load will then hold the voltage in balance (providing there is enough load to place the transmission voltage within its operating limits).

Figure 4-43 shows a 345 kV, 35 MVAR, three-phase shunt reactor used to help regulate transmission voltage during light load conditions and during the energization of long transmission lines.

Series Reactors–Distribution

Distribution substations occasionally use *series reactors* to reduce available fault current. Distribution lines connected to substations that have several transmission lines or are near a generation plant might have extremely high short circuit fault current available if something were to happen out on the distribution line. By inserting a series reactor on each phase of each distribution line, the fault current decreases due to the fact that a magnetic field has to be developed before high currents flow through the reactor. Therefore, the breaker trips the distribution line before the current has a chance to rise to full magnitude. Otherwise, the high fault current could cause excessive damage to consumers' electrical equipment. Series reactors are shown in Figure 4-44.

Figure 4-43   345 kV reactor.

STATIC VAR COMPENSATORS

The *static VAR compensator* (SVC) is a device used on ac transmission systems to control power flow, improve transient stability on power grids, and reduce system losses (see Figure 4-45). The SVC regulates voltage at its terminals by controlling

Series reactors

Figure 4-44   Series reactors.

Figure 4-45   Static VAR compensator. Courtesy of Jeff Selman.

the amount of reactive power injected or absorbed from the power system. The SVC is made up of several shunt capacitors and inductors (i.e., reactors) and an electronic switching system that enable ramping up or down reactive power support. When system voltage is low, the SVC generates reactive power (i.e., SVC capacitive). When system voltage is high, the SVC absorbs reactive power (i.e., SVC inductive). The variation of reactive power is performed by switching three-phase capacitor banks and inductor banks connected on the secondary side of a large coupling transformer.

CONTROL BUILDING

Control buildings are commonly found in the larger substations. They are used to house the equipment associated with the monitoring, control, and protection of the substation equipment (i.e., transformers, lines, and bus). The control building as in Figure 4-46 contains protective relaying, breaker controls, metering, communications, batteries, and battery chargers.

The protective relays, metering equipment, and associated control switches are normally mounted on relay racks or panels inside the control building. These panels also include status indicators, sequence of events recorders (SOE), computer terminals for system control communications, and other equipment that requires environmental conditioning. CTs and PTs cables from the outside yard equipment also terminate in the control building in cabinets or relay panels.

The environmental conditioning in a control building usually consists of lighting, heating, and air conditioning to keep the electronic equipment operating reliably.

Control buildings house important *sequence of events recorders* (SOE) needed to accurately track the operation of all substation equipment activity, primarily just before, during and after system disturbances. Accurate time stamps are placed on each event for follow-up analysis. Some of the items tracked include relay operations

Figure 4-46   Control building. Courtesy of Schweitzer Engineering Laboratories, Inc.

and circuit breaker trip information. The recorder produces a data file or paper record of all events that occurred at that substation during a major disturbance. This information is later analyzed with SOE data from other substations (including those from other utilities) to determine what went right, wrong, and what changes are needed to avoid similar disturbances in the future. This information is time stamped with highly accurate satellite clocks. This enables one to analyze an interconnected power system disturbance to determine whether the equipment operated properly and what recommendations are needed.

Please note that "Substation Automation" or "Smart Substations" is discussed in Chapter 7 and Chapter 9.

PREVENTIVE MAINTENANCE

Electric power systems have many ways to perform preventive maintenance. Scheduled maintenance programs, site inspections, routine data collection, and analysis are very effective. An enhanced or a more effective means of performing preventive maintenance is *predictive maintenance*. Sometimes this is called "*condition based maintenance*" where maintenance is based on measured or calculated need rather than just a schedule. Predictive maintenance can identify potential serious problems before they occur. Two very effective predictive maintenance programs or procedures are *infrared scanning* and *dissolved gas analysis* testing.

Infrared Technology

Infrared technology has improved maintenance procedures significantly. Temperature sensitive cameras are used to identify hot spots or hot hardware. "Hot" in this case is referring to excessive heat opposed to hot referencing "energized" equipment. Loose connectors for example can show up on infrared scans very noticeably. Loose connections can be very hot due to the high resistance connection compared to the temperature of surrounding hardware, indicating a problem exists. Extreme hot spots must be dealt with immediately before failure occurs.

Infrared technology is a very effective *predictive maintenance* technique. Infrared scanning programs are used by most electric utilities. Scanning many types of equipment such as underground, overhead, transmission, distribution, substation, and consumer services is a cost effective means of preventive maintenance.

Dissolved Gas Analysis

Dissolved gas analysis (DGA) is another very effective predictive maintenance procedure that is used to determine the internal condition of a transformer. Taking small oil samples periodically from important transformers allows one to accurately track and, through trend analysis, determine if the transformer experienced arcs, overheating, corona, sparking, etc. These types of internal problems produce small levels of various gases in the oil. Specific gases are generated by the certain problem conditions. For example, if an oil analysis finds existence of abnormally high levels of carbon dioxide and carbon monoxide gases, this might indicate some overheating of the paper insulation used around copper wires in the transformer coils. Acetylene gas might indicate arcing has occurred inside the transformer.

Samples are taken periodically and the gas analysis compared to previous samples in a trend analysis procedure. Significant changes in the parts per million (PPM) values of the various gases could indicate problems exist inside the transformer. Critical transformers (i.e., generator step up or transmission transformers) might have equipment to continuously monitor or perhaps samples are taken every 6 months. Less critical transformers might have samples taken every year or two.

Once it has been determined that a transformer has a gas problem, it is immediately taken out of service and internally inspected. Sometimes the problems can be repaired in the field; for example, loose bushing/jumper connections causing overheating can be tightened. Sometimes the problem cannot be adequately determined in the field, in which case, the transformer has to be rebuilt. Repairing a transformer can be very costly and time consuming. However, it is much less costly to repair a transformer under controlled conditions than it is to face the consequences of a major transformer failure while in service.

CHAPTER 5

DISTRIBUTION

CHAPTER OBJECTIVES

After completing this chapter, the reader will be able to:

- ☑ *Explain the basic concepts of overhead and underground distribution systems*
- ☑ *Discuss how and why distribution feeders are operated radially*
- ☑ *Discuss the differences between grounded wye and delta distribution feeders and laterals*
- ☑ *Explain the differences between single-phase and three-phase services*

DISTRIBUTION SYSTEMS

Distribution systems are responsible for delivering electrical energy from the distribution substation, like that shown in Figure 5-1, to the service entrance equipment located at residential, commercial, and industrial consumer facilities. Most distribution systems in the United States operate at primary voltages between 12.5 kV and 24.9 kV. Some operate at 34.5 kV and some operate at lower distribution voltages such as 4 kV. These low-voltage distribution systems are being phased out because of their relatively high cost for losses (low voltage requires high currents, which means high losses).

Distribution transformers mounted on poles or underground padmounts near customer load centers convert the *primary* voltage to *secondary* consumer voltages. This chapter discusses distribution systems between the substation and consumer service transformer.

Distribution Voltages

Table 5-1 shows the various distribution system voltages used in North America. This table is not absolute; some power companies may designate their system voltages differently.

System voltage is a term used to identify whether the reference is being made to *secondary* or *primary* distribution systems. Residential, commercial, and

Electric Power System Basics for the Nonelectrical Professional, Second Edition. Steven W. Blume.
© 2017 by The Institute of Electrical and Electronics Engineers, Inc. Published 2017 by John Wiley & Sons, Inc.

Figure 5-1 Distribution substation. Reproduced with permission of Fotosearch.

small industrial loads are normally served with voltages under 600 V. Manufacturers have standardized on providing insulated wire with a maximum 600 Vac rating for "secondary" services. For example, household wires, such as extension cords, have a 600 Vac insulation rating. Other than changing the plugs and sockets on either end, one could use this wire for higher secondary voltages such as 240 Vac.

The 34.5 kV system voltage is used differently among electric power companies. Some companies use 34.5 kV distribution system voltages to connect service transformers in order to provide secondary voltages to consumers while other companies use 34.5 kV power lines between distribution substations as a "sub-transmission" and not for consumers.

TABLE 5-1 Common Distribution Voltages

System Voltage	Voltage Class	Nominal Voltage (kV)	Voltage Category
Secondary	Under 600	0.120/0.240/0.208	
0.277/0.480	Low voltage (LV)		
Distribution	601–7200	2.4–4.16	Medium voltage (MV)
	15,000	12.5–14.4	High voltage (HV)
	25,000	24.9	
Distribution or sub-transmission	34,500	34.5	

There are several common distribution system voltages used in the industry between secondary and 34.5 kV. For example, many power companies have standardized on 12.5 kV while others use 25 kV. Some companies use 13.2 kV, 13.8 kV, 14.4 kV, 20 kV, etc. There are several areas still using 4.16 kV systems. These lower-voltage distribution systems are quickly being phased out due to their high losses and short distance capabilities.

The *voltage category* for distribution is usually *high voltage*. Utilities often place HV warning signs on power poles and other associated electrical equipment as a safety precaution measure.

Distribution Feeders

Distribution lines or feeders like that shown in Figure 5-2 are normally connected *radially* out of the substation. Radially means that only one end of the distribution power line is connected to a source. Therefore, if the source end becomes opened (i.e., de-energized) the entire feeder is de-energized and all the consumers connected to that feeder are out of service.

The transmission side of the substation normally has multiple transmission lines feeding the substation. In this case the loss of a single transmission line should not de-energize the substation and all radial distribution feeders should still have a source of power to serve all consumers. The operative word is "should." Usually system protective relays control the switching operation of substation circuit breakers. There are rare occasions when the protective relaying equipment fails to perform as intended and inadvertent outage situations occur.

Distribution feeders might have several disconnect switches located throughout the line. These disconnect switches allow for load transfer capability among the feeders, isolation of line sections for maintenance, and visual openings for safety purposes while working on lines or equipment. Even though there might be several lines

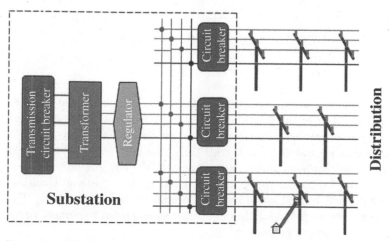

Figure 5-2   Distribution feeders.

and open/closed disconnect switches connected throughout a distribution system, the distribution lines are still operated or fed radially.

Wye vs. Delta Feeders and Connections

Similar to what was discussed earlier for connecting three-phase generators to transformers and transmission lines, distribution feeders also use wye and delta configurations. This section compares the two distribution feeder construction alternatives, wye (Figure 5-3) and delta (Figure 5-4). Most of the three-phase distribution feeders and transformer connections use the wye system alternative because of the advantages over the disadvantages. Although delta distribution systems do exist, much of the delta distribution has been converted to wye.

The *wye* and *delta configurations* are shown below:

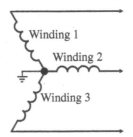

Figure 5-3   Wye connection.

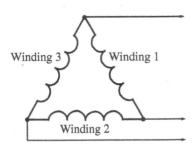

Figure 5-4   Delta connection.

The wye connected system has one wire from each coil connected together to form the *neutral*. Most of the time this neutral is *grounded*. Grounded implies that the three common wires are connected together and then connected to earth via a ground rod, primary neutral, or ground grid. The continually grounded neutral is also referred to as the "*multi-grounded neutral*" or "*MGN*". The grounding system provides a low resistance connection to earth. Grounding gives earth an electrical reference, hopefully 0 V. In the case of wye, this neutral reference is normally 0 V. (Chapter 10: Safety covers grounding in more detail and includes important safety aspects associated with proper grounding.)

The earth surface is conductive most of the time. Depending on the type of soil (rich fertile soil vs. granite rock) and the condition of the soil (wet vs. dry), earth can be a very good conductor or a very poor conductor (i.e., a very good insulator) or both depending on the season. Giving the distribution power line a reliable earth connection with the neutral zero voltage reference improves such things as safety, voltage stability, protection system design parameters, and other operating factors.

There are many applications that use the wye or delta configuration. Starting with distribution substations; the concept of wye or delta applies to the low-voltage side of the substation distribution transformer. (Most of them are grounded-wye connected.) Next is the distribution feeder being wye or delta. (Most are wye indicating a four-wire power line having "line-to–line" voltages and "line-to–neutral" voltages.) Each consumer that has a three-phase service transformer has to refer the high-voltage side of the transformer bank and the low-voltage side connections. (The common adopted standard is the wye-wye configuration distribution transformer bank.) Properly connecting three-phase loads to a distribution system involves knowing how the load equipment is supposed to be connected per the manufacturer (i.e., wye or delta) and how the distribution line is configured (i.e., wye or delta). There are multiple ways to configure three-phase equipment; however, the preferred means is to have four-wire wye equipment connected to four-wire wye-wye distribution transformers on a four-wire wye primary. This arrangement provides a highly preferred common neutral grounding system that offers safe and reliable service with minimal power quality issues.

Both wye and delta configurations have distinguishable advantages and disadvantages when it comes to transmission or distribution systems. Transmission and sub-transmission lines are built as three-phase, three-wire lines. The ends of the transmission lines are connected to either delta or *source grounded wye* transformer connections. "Source grounded wye" connection means that the transmission transformer in the substation is a four-wire wye transformer that has the three phases connected to the line conductors and the neutral connected to the substation *ground grid*. The neutral is not provided on transmission lines. It is not necessary to provide the neutral on transmission lines because all three phases are assumed to have balanced currents and there would not be current flowing in the neutral conductor when the currents are balanced. With respect to distribution lines, most systems use all four wires (including the neutral) as grounded-wye connections and current is usually present in the neutral because the three phase currents are normally not balanced. Distribution systems have single phase loads and therefore introduce neutral currents. At the substation however, hopefully the three-phase currents are balanced and the neutral current is zero leading up to the transmission source.

Three-wire delta distribution lines still exist in many areas. Although they are common in rural areas where a neutral is not present, delta distribution lines are also present in urban areas mainly in California and to a lesser degree in some mid-western states such as Illinois, Ohio, and Indiana. Upstate New York has a significant amount of delta distribution as do other identifiable spots in the United States. Those lines are more vulnerable to stray currents and voltages through the earth as the earth tries to handle the unbalanced current flow of distribution. The preferred standard for distribution is the grounded-wye configuration.

From the perspective of distribution systems, the following predominant advantages and disadvantages apply:

Advantages of Grounded—wye:

- Common ground reference

    Meaning: the power company's primary distribution neutral is grounded, the service transformer is grounded, and the customer's service entrance equipment is grounded all to the same reference voltage point.

- Better voltage stability

    Meaning: the common ground improves voltage stability because the reference point is consistent. This also improves power quality.

- Lower operating voltage

    Meaning: equipment is connected to the "line-to-neutral (L-N)" potential instead of the higher-voltage "line-to-line (L-L)" connection.

- Smaller equipment size

    Meaning: since the equipment is connected to a lower voltage (line-to-neutral opposed to line-to-line), bushings, gap spacing, and insulation requirements can all be smaller.

- Can use single bushing transformers

    Meaning: since one side of the transformer winding is connected to the grounded neutral, that connection does not need a bushing. Instead, single bushing transformers have an internal connection to the neutral, thus saving money.

- Easier to detect line-to-ground faults

    Meaning: should a phase conductor fall to the ground, or a tree make contact with a phase conductor, etc. the resulting short fault circuit flows through the neutral back at the substation along with earth return current. Therefore, by simply measuring the substation transformer's neutral overcurrent condition using a current transformer (CT), the ground fault protective relay can sense the line-to-ground fault out on the line. This overcurrent condition causes the protective relay to initiate a trip signal to the corresponding feeder circuit breaker. (Note, in the case of delta feeder configurations, there is no grounded neutral and is thus much harder to detect line-to-ground faults.)

- Better single-phase protection with fuses

    Meaning: fuses on transformers and distribution feeder lateral extensions clear faults much more reliably than fuses connected in delta configurations. Since deltas have equipment connected line-to-line, a line-to-ground fault could blow one or more fuses. A line-to-ground fault on a delta feeder may fatigue or weaken fuses on other phases. Therefore, it is a common practice on delta systems to replace all three fuses in case one or more were weakened from a line-to-ground fault.

Disadvantages of Grounded—wye:

- Requires four conductors

   Delta systems require only three conductors for three-phase power. The early days of electrifying America used this three-conductor approach in distribution feeders for its cost savings advantage, however it was later reconsidered by most utilities to add the fourth wire and take advantage of the several more benefits obtained by using the four conductor approach. Today, most of these lines have been converted to four-wire wye systems for the advantages that wye has over delta.

Advantages of Delta

- Three conductors versus four (i.e., less expensive to construct)
- Power quality enhancement

   The third-order harmonics are eliminated due to a natural cancellation. In other words, the 60 Hz power sine wave is cleaner by nature. The 120 degree phase shift between phases acts to cancel out some unwanted interference voltages.

- Lightning performance

   One could argue that sometimes the isolated conductors in a delta from ground reference minimize the effect that lightning has on a system. However, lightning arresters in delta systems are still connected line to earth ground.

Disadvantages of Delta

- No ground reference

   Meaning; service voltage may be less stable, fuse protection may be less effective, and there might be more overall power quality issues.

- Stray currents

   Distribution transformers can cause stray currents to flow in the earth when their low-voltage secondary side is grounded. Although the primary side of the distribution transformer is not grounded, the secondary side is grounded. Therefore, a small but measurable voltage is inadvertently connected to ground causing stray currents to exist. This situation gets worse when there are nearby lightning strikes and power faults.

- Unbalanced currents

   Three-phase transformer banks can regulate the primary voltage or try to equalize the primary voltage. The delta connections on the secondary side of the service transformer can cause the primary voltage to equalize. This can result in additional stray currents or unbalanced currents in the feeder. This situation gets worse as the distance between the substation and service transformer increases.

   Fault locating is more complicated or complex due to the lack of a neutral conductor. Oftentimes you have to wait until there is a two-phase fault to detect that there is a problem.

Comparing all the advantages and disadvantages, multi-grounded neutral, four-wire wye distribution feeders is the preferred method. Most of the delta distribution lines have been replaced with grounded-wye systems, however, some deltas do exist. The preference is for power companies to use grounded-wye systems on all distribution systems. Converting delta to wye does require the cost of adding a conductor. Conversion can be a slow process and is usually done over time as load growth occurs in the delta areas.

Line-to-Ground vs. Line-to-Neutral Voltages *(Optional Supplementary Reading)*

Grounded-wye systems have two voltages available for use. These two voltages are related mathematically by the $\sqrt{3}$. Equipment can be connected either "line-to-line" (L-L) or "line-to-neutral" (L-N). The L-N voltage is less than the L-L voltage by the $\sqrt{3}$. The neutral side of the L-N voltage is normally connected to earth by means of ground rods or grounding wires. The lower-voltage L-N connection is the normally used connection. Therefore, distribution power is transported efficiently in wye distribution systems yet consumer transformers are connected to a lower L-N voltage source. The intent is to balance all the L-N connections such that the load, as seen by the substation transformer, is balanced among all three phases.

For example, 12.5 kV L-L distribution systems have a 7.2 kV L-N voltage available for transformer connections. (Hence, 12.5 kV divided by $\sqrt{3}$ equals 7.2 kV)

Lastly, the term "line" is interchangeable with the term "phase". It is correct to say either "line-to-line" or "phase-to-phase". It is also acceptable to say the terms "line-to-neutral" or "phase-to-neutral".

Wye Primaries Overhead

WYE connected primary distribution lines consist of three phases and a neutral, as shown in Figures 5-5 and 5-6. The neutral is grounded at every pole in most systems (note, some rural grounded-wye systems might have a local practice of grounding a minimum of five grounds per mile and not every pole). Whereas, the National Electrical Safety Code (NESC) requires 4 grounds per mile and considers this a *multigrounded neutral* (MGN). One can identify a wye primary configuration by the way single-phase transformers are connected to the line. One of the transformer bushings will be grounded. Examining how the wires are connected to the transformer bushings helps one determine if the transformer is connected line-to-ground or line-to-line. Some wye connected transformers only have one high-voltage bushing. In that case, the neutral side of the primary winding is internally connected to the tank ground lug that is connected to the primary neutral. (Special note, single-phase transformers can be connected line-to-line as well as line-to-neutral, provided they have two bushings. They do not have to be connected line-to-neutral. Actually, this is a common occurrence where a delta distribution line is converted to a grounded wye line. The fourth conductor (neutral) is added to the three conductor delta. The old line-to-line transformers are now connected line-to-neutral. Therefore, the line-to-line voltage was increased by $\sqrt{3}$. This raised the feeder voltage, which in turn reduces system losses.]

Figure 5-5   Wye distribution.

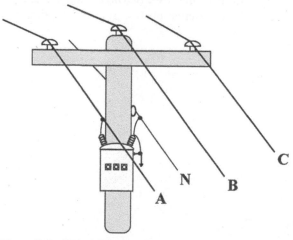

Figure 5-6   Wye three-phase feeder. Courtesy of Alliant Energy.

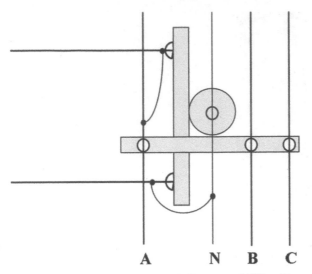

Figure 5-7 Wye one-phase lateral. Courtesy of Alliant Energy.

Lateral single-phase feeders branching off wye primaries usually consist of one-phase conductor and a neutral conductor as shown in Figure 5-7. At the branch point, the neutral will be grounded and follow the company's neutral grounding practice along the lateral. Figure 5-8 shows a transformer connected to a single-phase lateral.

Delta Primaries

Delta primary distribution lines use three conductors (one for each phase) and no neutral (see Figure 5-9). Single-phase transformers must have two high-voltage bushings and each bushing must connect directly to different phases. Since delta primaries

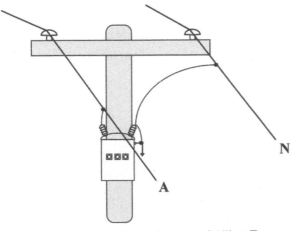

Figure 5-8 One-phase lateral. Courtesy of Alliant Energy.

Figure 5-9   Delta distribution.

do not have primary neutrals, the transformer tank grounds and lightning arrester grounds must be connected to a ground rod at the base of the pole with a ground wire along the side of the pole. Delta primaries and fused laterals require single-phase transformers to be connected phase-to-phase. Figures 5-10 to 5-12 show delta primary distribution lines.

The single-phase delta laterals consist of two-phase conductors and no neutral.

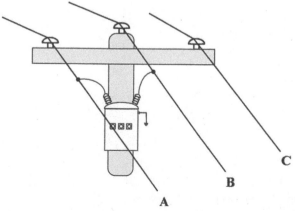

Figure 5-10   Delta three-phase feeder. Courtesy of Alliant Energy.

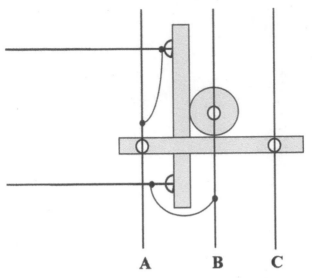

Figure 5-11   Delta one-phase lateral. Courtesy of Alliant Energy.

TRANSFORMER CONNECTIONS (OPTIONAL SUPPLEMENTARY READING)

This section discusses the most *common transformer configurations*; phase-to-neutral (i.e., line-to-ground) for single-phase connections and wye-wye for three-phase transformer bank connections. The most common connection for a distribution transformer, single phase or three phase, is phase-to-ground (i.e., line-to-ground). Figure 5-13 shows a one-phase transformer installation.

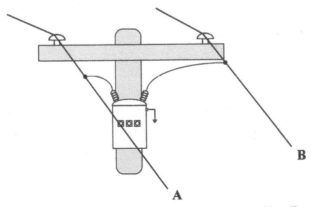

Figure 5-12   Delta one-phase lateral. Courtesy of Alliant Energy.

Figure 5-13   Transformer connections.

Distribution Transformers: Single-Phase

Since the standard residential service voltage is 120/240 Vac, most distribution transformers have turns ratios that produce the 120/240 Vac on their secondary or low-voltage side. Service wires are connected between the distribution transformer secondary side bushings and the consumer's service entrance equipment.

Transformer Secondary Connections: Residential

In order to produce the two 120 Vac sources (to make up the 120 Vac and the 240 Vac service) for the *residential consumer*, the distribution transformer has two secondary windings.

Figure 5-14 shows how the 120 Vac and the 240 Vac service is provided from the secondary side of a single distribution transformer. Figure 5-15 shows the transformer connections. This is the most standard connection configuration for residential consumers. This single-phase transformer has the two 120 Vac low-side voltage terminals connected in series with a neutral connection in the middle. This transformer supplies 120/240 V single-phase service to residential customers. Note the two secondary windings in series.

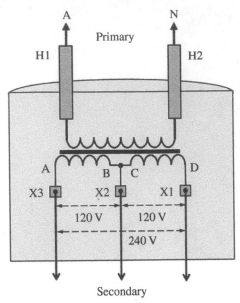

Figure 5-14   Standard two-bushing transformer.

Note the bushing nomenclature. The H1 and H2 markings identify the high-voltage side connections (i.e., bushings). The X1 and X2 identifies the low-voltage side connections (i.e., bushings). This is common nomenclature practice for all voltage classes including very high voltage transformers.

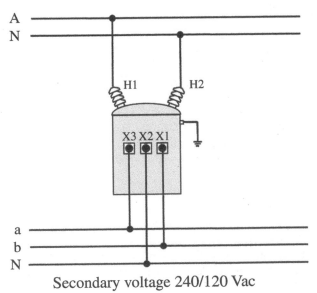

Secondary voltage 240/120 Vac

Figure 5-15   Two-bushing transformer connections.

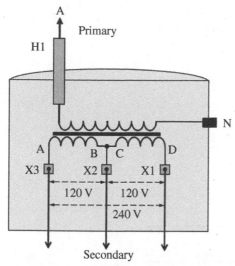

Figure 5-16   Standard one-bushing transformer.

Example, suppose the distribution feeder voltages were 12.5 kV line-to-line which has a line-to-neutral voltage of 7.2 kV (hence, divide the line-to-line voltage by the square root of three). Using transformers with 60:1 turns ratios on each of the two secondary windings, the secondary voltage becomes 120 V (7200 V divided by 60). The two secondary windings together produce 240 V.

Single-Phase One-Bushing Transformer
Figure 5-16 is also a single-phase transformer, however one high side bushing has been eliminated. Since one side of the primary winding is connected to neutral anyway (see Figure 5-17), the connection is made internally. This is referred to as a *single bushing transformer*. It has a terminal lug for the neutral or ground connection. In some transformers, the X2 bushing is also internally connected to the ground connection.

Distribution Transformers: Three-Phase

Three single-phase transformers are used to produce three-phase service to *commercial and industrial consumers*. The small commercial and light industrial consumers are normally served with 208/120 Vac three-phase service. The larger commercial and industrial consumers are normally served with 480/277 Vac three-phase service. This section discusses how the three-phase service voltages are produced. Figure 5-18 shows a typical three-phase 208/120 Vac transformer bank.

Transformer Internal Connections
Standard single-phase distribution transformers must be modified internally to produce only 120 Vac opposed to 120/240 Vac if they are to be used in *three-phase*

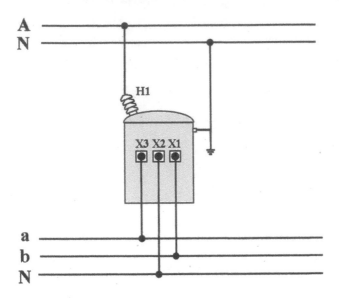

Secondary voltage 240/120 Vac

Figure 5-17   One-bushing transformer connections.

Figure 5-18   Three-phase transformer bank.

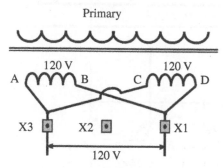

Figure 5-19   Transformer bank connection #1.

transformer banks. Two of the possible three ways to internally connect the two secondary windings to produce only 120 Vac are shown in Figures 5-19 and 5-20. These transformers supply 120 V only. The reason Figure 5-20 would be preferred by a power company is its similarity to connecting a standard 120/240 transformer, in that the center secondary neutral bushing connection (neutral) is in the same position in both cases.

The Standard Three-Phase Wye-Wye Transformer Bank (208/120 Vac)
Three single-phase transformers are connected together to form a transformer bank. The most popular three-phase transformer bank configuration (i.e., wye-wye) is shown in Figure 5-21.

The Standard Three-Phase Wye-Wye Transformer Bank (480/277 Vac)
Larger consumers require 480/277 Vac three-phase power. The three-phase transformer configuration that follows current standards is the three-phase 480/277 Vac wye-wye as shown in Figure 5-22. *Industrial consumers* that have large motors, several story buildings, several lights, elevators, etc. usually require the higher 480/277 Vac service opposed to the lower 208/120 Vac service. (Note, again the higher-voltage system requires lower currents, smaller wires, less losses, etc. for the same power

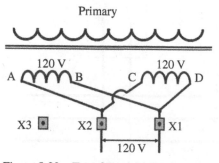

Figure 5-20   Transformer bank connection #2.

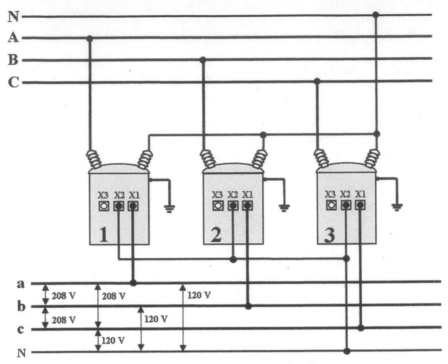

Figure 5-21   208/120 Vac, three-phase wye-wye connection diagram.

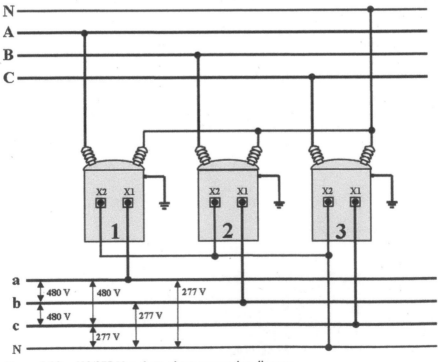

Figure 5-22   480/277 Vac, three-phase connection diagram.

level. Therefore, wire capacity is made available due to the higher voltage and less current.)

Delta Connections

Delta-delta, wye-delta, and delta-wye distribution transformer bank configurations are not as common as the standard wye-wye configuration and therefore are not discussed in this book.

Dry-Pack Transformers

Consumers that take service at 480/277 Vac usually require *dry-pack* transformers at their facility to provide 208/120 Vac service to power standard receptacles and other basic 120 Vac necessities. Dry pack implies no insulation oil is contained in the transformer. These dry-pack transformers are often located in closets or small rooms with high-voltage warning signs posted on the door. Figure 5-23 is an example of a dry-pack transformer.

Most of the *large motor loads* (i.e., elevators) at these larger consumers operate at 480 Vac three-phase. The large arrays of lighting use 277 Vac line-to-ground single-phase power. Therefore, just the basic 120 V loads use the dry-pack transformers.

Note, 277 Vac line-to-neutral is a very common industrial voltage for many motors, lights, etc. 277 V equipment might not be found at your local department

Figure 5-23   Dry-pack transformer. Courtesy of Alliant Energy.

stores, but industrial electrical supply stores are packed full with 277 V equipment to be installed at industrial facilities.

FUSES AND CUTOUTS

The purposes of a *fuse* are to interrupt power flowing to equipment when excessive current occurs and to provide equipment damage protection due to short circuits and power faults. Fuses like that in Figure 5-24 interrupt the flow of current when the maximum continuous current rating of the fuse is exceeded. The fuse takes a very short period of time to melt open when the current rating is exceeded. The higher the excessive current, the faster the fuse melts.

   Fused cutouts like that shown in Figure 5-25 are the most common means of protection devices in the distribution system. They are used to protect distribution transformers, underground feeds, capacitor banks, PTs, substation service power, and other equipment. When blown, the fused cutout door falls open and provides a visible break in the circuit for line workers to see. The hinged door falls open and hangs downward as shown in Figure 5-26. Sometimes the door does not fall open; it remains intact due to ice, corrosion from salt fog, or other mechanical operation infringements. Line workers on patrol for an outage can normally see the blown fuse from a distance.

   Comparing fuses to circuit breakers, circuit breakers have the ability to open and close circuits repeatedly while a fuse opens the circuit one time and must be replaced. Fuses are single-phase devices, whereas circuit breakers are normally gang operated three-phase devices. Breakers can interrupt very high magnitudes of current.

Figure 5-24   Distribution fuse.

Figure 5-25   Fuse cutout. Courtesy of Alliant Energy.

Figure 5-26   Fuse door. Courtesy of Alliant Energy.

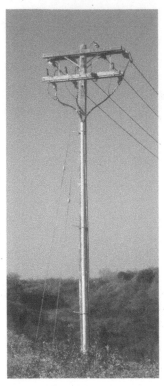

Figure 5-27   Dip pole or riser pole.

Breakers close into a fault and trip open again. Breakers can be controlled remotely. And, breakers need periodic maintenance.

RISER OR DIP POLE

The purpose of a riser or dip pole is to transition from overhead construction to underground construction. Some electric utilities refer to them as *dip poles* when the power source is the overhead serving the underground and *riser poles* when the source is underground serving the overhead. Either way, they represent an overhead to underground transition. An example of a typical transition pole is shown in Figure 5-27.

UNDERGROUND SERVICE

Underground construction is usually about 3–5 times more costly than overhead construction. Most people prefer underground construction as opposed to overhead for aesthetic reasons. Underground systems are not exposed to birds, trees, wind, lightning etc., and should be more reliable. However, underground systems fault due to

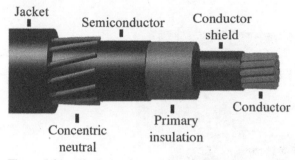

Figure 5-28　Single conductor primary distribution cable. Courtesy of Alliant Energy.

cable, elbow, splice, dig-in, and connector failures. When underground systems fault, they usually cause significant damage (i.e., cable, elbow, or splice failure). Therefore, underground feeders are usually not automatically reclosed.

Primary Distribution Cable

Primary underground cables are one of the most important parts of any underground system. If a fault occurs on an underground cable, the feeder or fused section of line is out of service until a crew can isolate the bad section of cable and perform necessary load transfer switching to restore power. Note distribution automation (discussed later) would do this activity automatically, thus significantly reducing outage time.

Most primary distribution cables like the one shown in Figure 5-28 consist of two conductors (*main center conductor* and *concentric neutral* conductor) with layers of insulation and semi-conductive wraps. The main center conductor is composed of either copper or aluminum. The outer conductor is the concentric neutral and is usually copper. The outer cover *jacket* is made of polyethylene, polyvinyl chloride (PVC), or thermoplastic material.

The concentric neutral helps trip a circuit breaker or fuse quickly if dug into by a backhoe or other type of dig in. Should a backhoe operator penetrate the cable, the blade is first grounded by the concentric neutral before striking the energized center conductor. This allows short circuit current to flow and trip the breaker without harming the equipment operator.

Underground cables have a significant amount of capacitance. When cables are de-energized, they can maintain or store a dangerous voltage charge. Special safety procedures are required when working with de-energized underground cables because of the stored or trapped voltage charge that can be present.

Load Break Elbow

Load break elbows are used to connect underground cables to transformers, switches, and other cabinet devices. As the name implies, load break elbows are designed to

Figure 5-29   Load break connections. Courtesy of Alliant Energy.

connect and disconnect energized lines to equipment. One can also energize and de-energize underground cables with load break elbows. However, safe working practices normally require that operating personnel use insulated rubber protection and fiberglass tools to insure safe working conditions are used when installing or removing elbows. Operators normally de-energize the equipment before connecting or removing load break devices. Figure 5-29 shows a line worker wearing rubber gloves removing an underground cable elbow using a fiberglass insulated tool.

Figures 5-30 and 5-31 show typical load break elbow connectors.

Splice

Underground splices are used to connect cable ends together. They are normally used for extending cable or emergency repairs. It is preferable not to use splices. They, like anything else add an element of exposure to failure.

Figures 5-32 and 5-33 show typical splices used in underground distribution systems. Note, all underground connections, especially elbows and splices require special installation procedures to assure high quality results for long-term reliable performance. Underground equipment is susceptible to water and rodent damage; therefore, extreme care must be taken when performing cable splicing.

Underground Single-Phase Standard Connection

Figure 5-34 shows a 7.2 kV-120/240 Vac, 25 kVA *single phase padmount trans-former*. Two high-voltage bushings are on the left and the low-voltage connectors

Figure 5-30   Load break elbow.

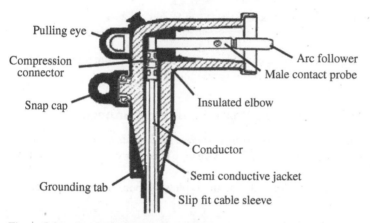

Pulling eye

Compression
connector

Snap cap

Arc follower

Male contact probe

Insulated elbow

Conductor

Semi conductive jacket

Grounding tab

Slip fit cable sleeve

Figure 5-31   Load break elbow components.

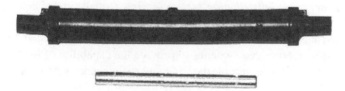

Figure 5-32   Underground long compression splice with cover. Courtesy of Alliant Energy.

Neutral   Ground
connection   connection

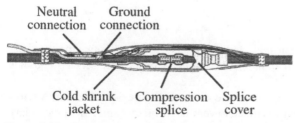

Cold shrink   Compression   Splice
jacket   splice   cover

Figure 5-33   3M primary underground splice. Courtesy of Alliant Energy.

Figure 5-34   Switching transformer. Courtesy of Alliant Energy.

are on the right. The two high-voltage bushings allow *daisy chaining* transformers in series to serve multiple residences in a loop arrangement.

Underground Wye-Wye Three-Phase Standard Connections

Figure 5-35 shows how an underground *three-phase padmount transformer* is connected to a four-wire wye primary and a four-wire wye secondary. This is very similar to the overhead wye-wye configuration.

This connection supplies 208/120 Vac three-phase service to the consumer.

Single-Phase Open Loop Underground System

A typical *single-phase underground distribution system* serving a small subdivision is shown in Figure 5-36. It provides reliable loop operation to several consumers. Notice the normally closed and open switches. This loop design uses "*daisy-chained*" padmount transformers with incorporated switches to provide the capability of load transfer and equipment isolation during troubleshooting and maintenance activities. Configurations like this allow a faulted section of cable to be isolated quickly and service restored while the cable is being repaired or replaced.

Secondary Service Wire

The electric utility is responsible for the service wire between the secondary side of the distribution transformer and the consumers' service entrance equipment.

Examples of secondary service wire are shown in Figure 5-37. Secondary wires are insulated. The insulation value is much lower than primary cable. Most secondary distribution wires consist of two insulated conductors and a neutral. Overhead service

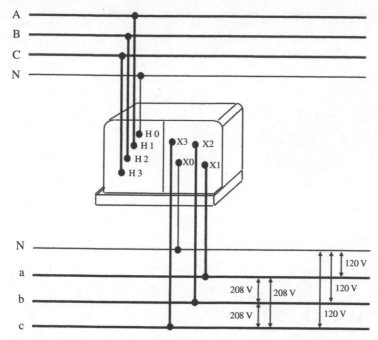

Figure 5-35   Underground wye-wye connection diagram.

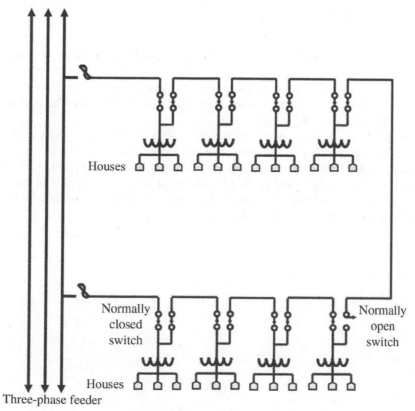

Figure 5-36   Distribution primary loop.

Figure 5-37   Secondary cable.

wires normally have the neutral conductor bare while underground service wires are all insulated.

The conductor's insulation is either polyethylene or rubber coated and is usually rated 600 Vac. The conductors are usually aluminum or copper. The neutral is usually the same size as the hot conductor.

Examples of overhead and underground *Triplex* cables are pictured in Figure 5-37. (Note, *Quadraplex* cables are used for three-phase services). To reduce the clearance needed for conductors from the service pole to the service entrance, conductors may be insulated and twisted together with a neutral. For single-phase service, such as street lighting, one insulated conductor is twisted together with an uninsulated neutral and referred to as *duplex* cable.

CHAPTER 6

CONSUMPTION

CHAPTER OBJECTIVES

After completing this chapter, the reader will be able to:

- ☑ *Explain the different categories of energy consumption (residential, commercial, and industrial) and their characteristics*
- ☑ *Explain how power factor and system efficiency are related*
- ☑ *Discuss metering, demand side management, and smart consumption*
- ☑ *Explain how to wire residential panels, lighting, receptacles, GFCI circuit breakers, and 240 Vac circuits*
- ☑ *Explain the common problems associated with starting large motors and how the various types of soft start equipment help reduce flicker*
- ☑ *Describe the types of power quality issues and resolutions*
- ☑ *Describe how industrial service entrance equipment, emergency generators, and uninterruptible power supply (UPS) systems work*

ELECTRICAL ENERGY CONSUMPTION

Consumption is the electrical energy used by all the various loads on the power system. Consumption also includes the energy used to transport and deliver the energy. For example, the losses due to heating conductors in power lines, transformers, etc. is considered consumption.

Electricity is consumed and measured several different ways depending on whether the load is residential, commercial, or industrial and whether the load is resistive, inductive, and capacitive. Electric utilities consume electricity just to produce and transport it to consumers. In all cases, electrical energy production and consumption is measured and accounted for. The electrical energy produced must equal the electrical energy consumed. This chapter discusses the consumption side of electric power systems. It also explores the types of load, their associated power requirements, and how system efficiency is measured and maintained.

In *residential* electric consumption, the larger users of electrical energy are items such as air-conditioning units, refrigerators, stoves, space heating, electric

Electric Power System Basics for the Nonelectrical Professional, Second Edition. Steven W. Blume.
© 2017 by The Institute of Electrical and Electronics Engineers, Inc. Published 2017 by John Wiley & Sons, Inc.

water heaters, clothes dryers and to a lesser degree lighting, radios, and TVs. Typically, all other home appliances and home office equipment use less energy and therefore account for a small percentage of total residential consumption. Residential consumption has steadily grown over the years and appears that this trend is continuing. Residential energy consumption is measured in kilowatt-hours (kWh).

Commercial electric consumption is also steadily growing. Commercial loads include mercantile and service, office operations, warehousing and storage, education, public assembly, lodging, health care, and food sales and services. Commercial consumption includes larger scale lighting, heating, air conditioning, kitchen apparatus, and motor loads such as elevators and large clothes handling equipment. Typically, special metering is used to *record peak demand* (in kilowatts) along with energy consumption in kWh.

Industrial electric consumption appears to be steady. Industrial loads usually involve large motors, heavy duty machinery, high-volume air-conditioning systems, etc. where special metering equipment is used such as *power factor, demand, and energy*. Normally the consumption is great enough to use CTs (current transformers) and PTs (potential transformers) to scale down the electrical quantities for standard metering equipment.

Primary metering is used for very large electrical energy consumers (i.e., military bases, oil refineries, mining industry, etc.) to measure their consumption. These large consumers normally have their own sub-transmission and/or primary distribution facilities including substations, lines, and electrical protection equipment.

Consumption Characteristics

The combination of the three types of load (i.e., resistors, inductors, and capacitors) working together in power systems influences system losses, voltage stability, revenues, and reliability. This section explains how these loads interact and how their interaction improves (or impedes) the overall performance of electrical power system operations.

Basic ac Circuits
The three basic types of ac circuits are Resistive, Inductive, and Capacitive. These circuit types with ac power sources are shown in Figure 6-1.

Depending on the type of load connected to an ac voltage source, a time angle difference between the voltage and current exists. This time difference is also referred

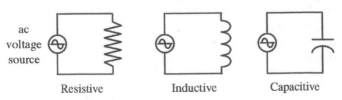

Figure 6-1   Types of circuits.

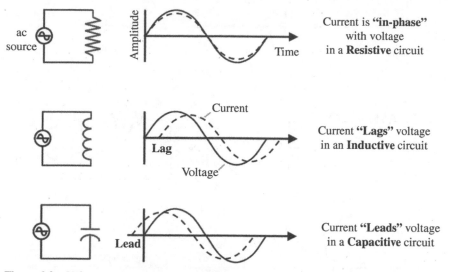

Current is **"in-phase"**
with voltage
in a **Resistive** circuit

Current **"Lags"** voltage
in an **Inductive** circuit

Current **"Leads"** voltage
in a **Capacitive** circuit

Figure 6-2   Voltage and current relationships.

to as the *phase angle* (lead or lag) between the voltage and current. The phase angle is usually measured in degrees since there are 360 degrees in one complete cycle.

Phase Angle Comparisons Between Load Types
The phase angle between the voltage and current is different for each basic load type. Figure 6-2 shows the three load type phase angles.

Combining Load Types
When both inductive loads and capacitor loads are connected together, their phase angles combine. Figure 6-3 shows this concept. *Part A* shows how the lagging inductive phase angle added to the leading capacitive phase angle combine and becomes equal to the phase angle of resistive loads. The phase angle does not have to cancel entirely; the net result can be either inductive lagging or capacitive leading. *Part B* shows two ways capacitors and inductors can be connected together to make their phase angles combine. *Part C* shows the equivalent resistive circuit when combining these two electrical components at full cancellation.

POWER SYSTEM EFFICIENCY

The *efficiency* of a power system is maximized when the total combined load is purely resistive. Therefore, when the total load on the system approaches purely resistive, the total current requirements and losses are minimum. The total power that has to be produced is minimized when the load is purely resistive. The total power becomes "real" power (i.e., Watt power) only.

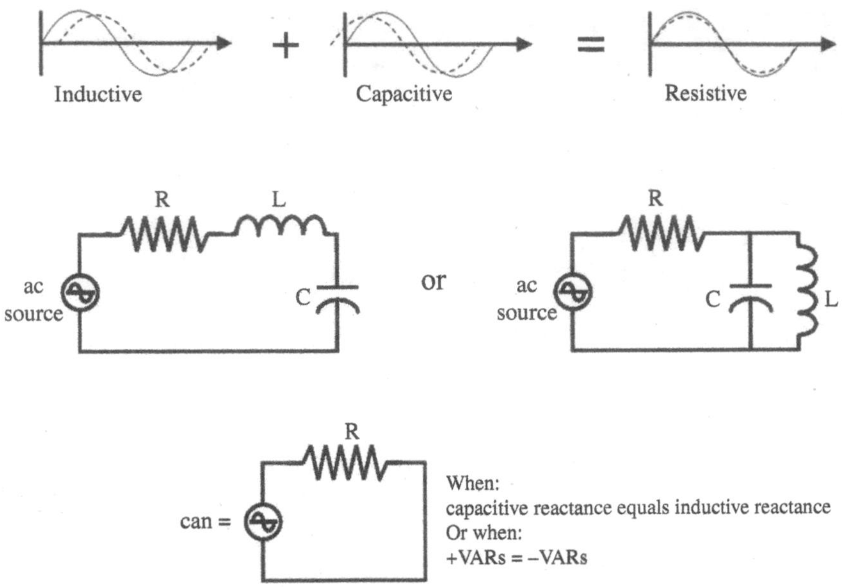

Figure 6-3   Equivalent circuits.

When the system efficiency is maximized (i.e., minimum power required to serve all loads), two significant benefits are realized.

1. Power losses are minimized
2. Extra capacity is made available in the transmission lines, distribution lines, transformers, and other substation equipment because this equipment is rated on the amount of total current carrying capability. If the current flow is less, the equipment has more capacity available to serve additional load.

One way to measure the power system efficiency is by calculating the power factor.

POWER FACTOR

The efficiency of a power system can be viewed as: how much total power (i.e., "real" power plus "reactive" power) is required to get the "real" work done. The *power factor* is a calculation that is based on the ratio between real power and total power, as shown below:

$$\text{Power factor} = \frac{\text{Real power}}{\text{Total power}} \times 100\%$$

Typically, power factors above 95% are considered "good" (i.e., high) and power factors below 90% are considered "poor" (i.e., low). Some motors, for example, operate in the low 80–85% power factor range and the addition of capacitors would improve the overall efficiency of the power supply to the motor.

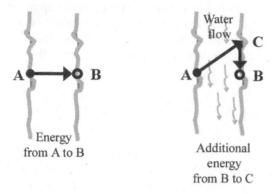

Figure 6-4   Power factor.

For example, suppose you were trying to cross a river from point "A" to point "B" as shown in Figure 6-4. The shortest path requiring the least amount of energy would be to swim a straight line, as shown on the left image. However, suppose water is flowing downward, causing you to swim a little upstream toward point "C" in order to arrive at point "B." The extra energy exerted from "C" to "B" would be considered wasteful. In electrical circuits, this wasteful opposing energy is called "reactive energy." Keep in mind that reactive power is needed for motors and other inductive load to produce their magnetic fields. Adding capacitors at the terminals of the motor to compensate for the reactive power needed by the motor means that reactive power is no longer needed by the generator and overall system losses are reduced.

(Optional Supplementary Reading)
Refer to Appendix B on a graphical analysis of power factor.

SUPPLY AND DEMAND

Let us put the power system into proper perspective; first comes voltage, then load, then current, then power, and then energy. Meanwhile, system losses occur requiring increased supply to achieve balance between generation (supply) and consumption (demand). During this process, the electric utilities try to maintain good regulated voltage and frequency for all consumption types and levels. The consumers draw the current, use the power and energy from the system. The consumers and system losses dictate the demand. The power producers must supply this demand through transmission and distribution systems to the consumers with good voltage and frequency. And, all of this is done in "real time."

Electric power systems operate in real time. As load increases, generation must increase to supply the demand or frequency drops. Also, as load increases, generation and line treatment (switched capacitors) must provide good voltage. Otherwise, voltage would collapse, frequency would drop, consumer lights would dim, and motors would overheat because load equipment is designed for a given voltage and frequency.

From a generator perspective, to increase frequency the prime mover spinning the turbine/generator rotor must increase. To increase voltage, the generator's exciter (the field current on the generator's rotor that makes the magnetic field) must increase. Voltage can also be increased by switching on shunt capacitors or switching off shunt reactors.

DEMAND SIDE MANAGEMENT

Since the amount of load (i.e., demand) determines the amount of generation (i.e., supply), the best way to minimize the need for additional supply is to reduce or control demand, especially during peak times. Demand side management (DSM) programs are being implemented everywhere to help manage load growth and achieve economic stability. *Demand side management* programs are designed to provide assistance to consumers in order to help reduce their energy demand, control their energy cost while delaying the construction of new generation, transmission, and distribution facilities. These DSM programs provide assistance by conducting energy audits, controlling consumer equipment (air conditions, heating, etc.), or by providing economic incentives (i.e., time of use rates). These programs are designed for residential and business consumers.

The kinds of incentives provided depend on the consumer type as described below:

Residential

The DSM programs that pertain primarily to residential and small business loads include the following:

- ➤ Lighting (i.e., rebate coupons, discounts for high efficiency light bulbs, efficient lighting designs, and other energy reduction incentives)
- ➤ High efficiency washing machines, clothes dryers, and refrigerators (rebates)
- ➤ Home energy audits (to identify usage patterns that can be improved)
- ➤ Insulation upgrades
- ➤ Appliance management (or smart control)
- ➤ Control some equipment to only operate during off-peak periods (water heaters, pool pumps, irrigation pumps, etc.)

Commercial

The DSM programs that involve commercial consumers are geared more toward overall operations efficiency, for example:

- ➤ The efficient design of buildings, remodeling or renovation activities using more energy efficient products, and technologies without increasing project costs. This would include; more efficient lighting, heating, air conditioning, motor upgrades, variable speed drives, and other more efficient electrical equipment.

➤ Replacement incentives to remove older, lower efficiency equipment

➤ Energy consumption analysis programs to encourage better operational methods within a business or organization, such as power factor correction capacitors.

Industrial

The DSM programs for industrial consumers focus on energy initiatives, for example,

➤ Renewable energy resources incentive programs to increase the utilization of wind power, solar energy, fuel cells, etc. to generate electricity for their own facility.

➤ Incorporation of online energy load profiles and real-time energy rate information are used to strategize efficient load patterns and conservation.

➤ Energy consumption surveys or studies to provide recommendations for load curtailment, cogenerations using standby emergency generators during peak demand.

There are other demand side incentives that are available to help reduce electrical energy consumption such as exterior or interior shading, awnings, wall glazing, heat reflectors, automatic control devices, and home or building energy management systems.

There is a concerted effort in the electrical industry to focus on ways to make electric energy consumption more efficient, less demanding, and less dependent on foreign energy resources. Energy production, transmission, and distribution costs, especially operational losses, expansion, and fuel dependency originate with consumption. Consumption control through "demand side management" programs is one of the best ways to postpone new generation projects, maintain or lower electric utility costs, and conserve energy.

METERING

Electric metering is the process for direct measurement of energy consumption. The electric quantities being measured depends on the consumer type and level of consumption. Residential consumers are metered for *energy* in kilowatt-hours (kWh) consumption. Small commercial and light industrial might have a *demand* meter as part of their metering package. Large industrial consumers might have energy (kWh), demand (peak kW), and *power factor* metering (%PF). The largest consumers of electricity might receive their power at distribution primary, sub-transmission or transmission voltage levels where *primary metering* is required.

Figure 6-5   Electro-mechanical kWh meter.

Residential Metering

The most common type of electric meter is the kilowatt-hour meter, like those shown in Figures 6-5 and 6-6. These meters measure electrical energy. Energy is the product of power and time. Note, since "Watts" are measured only, total power is not measured (i.e., total power would include reactive power in VARs). Therefore, the units measured are *watt-hours*. For scaling purposes, *kilowatt-hours* (*kWh*) are used as the standard unit for measuring electrical energy for residential customers.

The older dial type kWh meters (Figure 6-5) measure the actual energy flow in the three-conductor service wires from the power utility's distribution transformer. The current flowing in both legs (i.e., each 120 Vac conductor) and the voltage between the two legs provide the necessary information to record residential energy consumption. Residential load connected to 240 Vac is also measured because its current also flows through the two hot leg service conductors.

The dials turn in ratios of 10:1. In other words, the dial on the right makes 10 turns before the next dial on its left moves one indication, and so on. The difference between dial readings is the energy consumption for that period. The electronic or solid-state meters (Figure 6-6) record additional information such as time of use and in most cases can remotely communicate information to other locations through telephone lines, radio signals, power lines, or to small hand-held recording units.

Figure 6-6   Solid-state kWh meter.

Demand Metering

Small commercial and light industrial loads might have *demand* meters incorporated in their electrical metering equipment. The customer is charged for the highest sustained 15 minute sliding peak usage within a billing period. This type of metering is called demand metering. Some clock-type energy meters have a *sweep demand arm*, which shows the maximum 15-minute demand for that billing cycle. Figure 6-7 shows the demand needle and scale. Figure 6-8 shows a traditional clock type demand

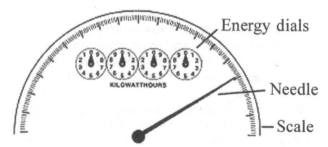

Figure 6-7   Demand needle and scale. Courtesy of Alliant Energy.

Figure 6-8   Demand meter. Courtesy of Alliant Energy.

and energy meter. Meter readers must manually reset the demand meter's sweep arm at each billing cycle. There are now electronic solid-state versions of demand meters in use. Some electronic meters electronically communicate this information to the electric utility, that is part of *advanced metering infrastructure* (AMR).

Time of Use Meters

A variation on demand metering is the *time of use* (TOU) metering. While demand meters measure the peak demand for each billing cycle, TOU meters record demand and energy consumption during the different parts of the day. TOU metering allows the utility to charge different rates for different parts of the day. For example, during the part of the day when energy consumption is highest with maximum generation online, or peak periods, TOU rates are higher than off-peak times. During the part of the day when energy consumption is lowest, or off-peak, TOU rates are much lower. These variable rates provide incentives to discourage consumption during peak hours and encourage consumption during off-peak hours. These units are almost always electronic solid-state meters with communications capabilities.

Smart Consumption

Smart meters, like the one shown in Figure 6-9 use two-way communications to help the power utility understand residential consumer's load characteristics, remote control service turn on/off, measure consumption based on time of use, and eventually more utilities will offer incentives to interconnect with the consumer's "home area networks" (HAN).

Home automation is available for the consumer to control consumption by controlling their own load such as room temperature, sprinkler systems, lighting, and smart appliances. The incentives are convenience while reducing the power bill and contributing toward reducing overall power grid consumption.

Figure 6-9   Residential smart meter.

Smart appliances available today enable consumers to optimize energy consumption, while at the same time providing convenience and accessibility. For example, smart thermostats in residential dwellings enable users to control their heating and air conditioning either remotely or from bed. The user can adjust climate as needed instead of fixed daily cycles. Kitchen automation enables users to remotely control, among other items, refrigerators, dishwashers, and stoves. Through mobile apps and appliance wireless internet connectivity, these smart appliances provide timely process control with progress updates, enabling the user to carry on normal life activities. Users can preheat, start, pause, stop, and adjust temperature settings and cook time, all from their smartphone. Smart dishwashers send alerts when a cycle is finished, when detergent replenishment is required, and can automatically reorder consumables. Display screens on refrigerator doors enable users to monitor status of all home smart devices, display calendar appointments, shopping lists, and do this through voice commands. The more users control their load efficiently through smart devices, the more grid energy is conserved and everyone benefits. Further, enabled utilities are able to control smart devices, especially high consumption devices such as water heaters and air conditions, when energy emergency alerts are declared.

Figure 6-10 shows how HANs can connect several ac and dc smart devices to create the ultimate residential electric power utilization system. This system incorporates the electric utility with home generation (e.g., rooftop solar), controllable load automation, electric vehicles, and much more. The future poses exciting new ways to reduce energy consumption through smart load and generation-enabled devices.

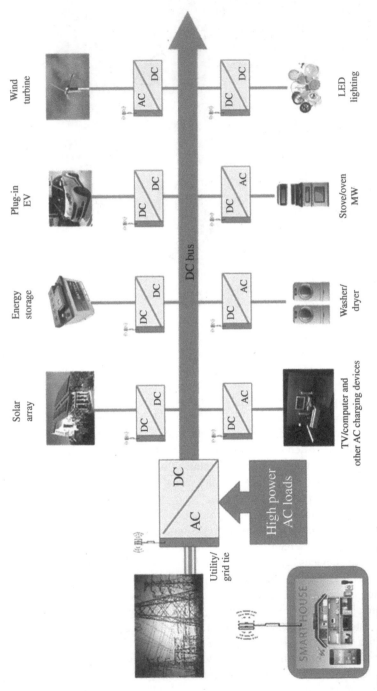

Figure 6-10 Home area network.

Reactive Meters

A watt-hour meter is neither designed nor intended to measure reactive power. However, by shifting the phase angle of the load CT (current transformer), a second watt-hour meter can be connected to this phase-shifted load that can measure reactive energy consumption. The phase is usually shifted with a capacitor–resistor network in single-phase systems; and with phase-shifting transformers in three-phase systems. The phase-shifting device helps measure the circuit's reactive power in *kilo-VAR-hours* (*kVAR*), or in units of 1000 VAR-hours. When connected this way, the second kVAR-hour meter is called a *reactive meter*. The electric utility can calculate the average power factor based on kWh and kVARh information. Some utilities employ direct reading power factor meters where peak power factor information is provided.

Primary Metering

Some customers have very large loads for their operation and require service at primary distribution voltage. Special primary voltage metering or metering at voltages above 600 V is required. Metering personnel install what is known as *primary metering* equipment, when it is not practical to do the metering at secondary voltage levels. Primary metering equipment includes high accuracy potential transformers (i.e., metering class PTs) and high accuracy current transformers (i.e., metering class CTs). Special structures, equipment cabinets, or equipment racks are required with this type of metering installation.

There are many possible ways to build primary metering equipment housings depending on the type of application. For example, primary metering equipment might apply to underground (Figure 6-11), overhead (Figure 6-12), substation, or industrial installations.

PERFORMANCE-BASED RATES

Some regulated utilities (i.e., distribution companies) are being faced with performance-based rates. That is, the Public Service Commission or other regulatory agency has established certain performance criteria relating to customer service reliability and a utility's reliability performance is taken into consideration when rate increases are requested. If the utility meets or exceeds set criteria, it is usually allowed to collect a "bonus" on the base rate. On the other hand, if the utility fails to meet the established criteria, the utility may be penalized with a lower rate of return. It behooves utilities (and customers) to improve their reliability performance.

Some of the performance-based rate indices include:

- ☑ SAIFI, which stands for system average interruption frequency index.
- ☑ SAMII, which stands for system average momentary interruption index.
- ☑ SAIDI, which stands for system average interruption duration index.

Figure 6-11   Underground primary metering. Courtesy of Alliant Energy.

All of these and other measurements focus on reliable service to the customer. These indexes were originally manually derived based on daily outage reports and are now supplied by system control center computers.

Reliability and stability of the overall power grid system that involves generation and transmission is discussed later in Chapter 8.

SERVICE ENTRANCE EQUIPMENT

The electric utility connects their service wires to the consumer's service entrance equipment. The National Electric Code (NEC) has very specific rules, regulations, and requirements on how service entrance equipment must be designed, installed, connected, and inspected. This section discusses the basic equipment designs, demand side connections, and special load characteristics considerations used for residential, commercial, and industrial consumers.

Residential Service Entrance Equipment

Actual service entrance equipment can vary from manufacturer to manufacturer, however the basic designs and concepts are standard. The concept is to provide a standard

Figure 6-12   Overhead primary metering.

and practical means of connecting the electric utility's 120/240 Vac single-phase service having two hot legs and one neutral wire to residential loads throughout the premises.

The standard *distribution service panel* is usually designed to encourage the balancing of the 120 V hot legs with connected loads. These designs usually make it convenient to connect 240 V loads with one combo circuit breaker. Since each consecutive circuit breaker space connects to opposite hot legs, any two adjacent breaker spaces conveniently connects both hot legs for 240 Vac applications. Therefore, a 240 Vac connection is accomplished by connecting two adjacent 120 V breakers and providing a plastic *bridge clip* across the two 120 V breakers causing both breakers to trip if a problem occurs on either hot leg.

A typical 120/240 Vac panel is shown in Figure 6-13, note the meter socket. Figure 6-14 shows the same residential panel with the outside cover removed, exposing the meter socket and breaker spaces. Figure 6-15 shows the panel with the breaker space cover removed. This panel is ready for wiring.

The left panel shows the location for the kWh meter. The center panel shows the meter socket connections. The right panel shows the individual breaker position spaces (without the breakers). Note the center of the lower portion has the metal tabs alternating left and right. This allows vertically adjacent circuit breakers to connect to opposing legs for 240 V service.

Figure 6-13   Basic panel.

Service Entrance Panel

The drawing in Figure 6-16 shows how the two hot legs and neutral are connected inside a typical distribution panel. The primary neutral is connected to the *neutral bus bar*. The service panel's *grounding bus bar* is connected to the grounding electrode system. A recent change to the NEC allows the use of concrete-encased electrodes (also known as Ufer grounds after its developer Herbert G. Ufer), often found in the foundation of a structure, as part of the grounding electrode system. This common ground connection improves voltage stability, protection equipment effectiveness, and safety. It is most effective with wye connected primary distribution systems opposed to delta-connected systems.

The two hot legs of the service entrance conductors are first connected to the *main breaker*. Notice how adjacent circuit breakers connect to opposite legs of the 120/240 Vac service wires. This arrangement encourages the balancing of load. Further, connections to two adjacent breakers provide the 240 Vac source. Note the two breakers involved in providing 240 V service has a plastic clip across their levers so that if one leg trips both legs trip.

Figure 6-14   Meter cover removed.

Light Switch

Figure 6-17 shows how a standard light switch circuit is configured. The NEC color code standard is stated.

Note how the wires can be extended to connect additional loads for a single breaker. The green ground wire is used to connect the light fixture metal to ground. Also note how the green ground wire and the white neutral wire eventually connect to the same ground at the service panel. The reason for connecting the two wires together is to provide an appliance ground connection, should a hot wire fray and short out to the metal appliance. The exposed hot wire shorts out with the metal appliance hot and ground wires and trips the panel breaker, thus removing a potentially dangerous situation.

Receptacle

Figure 6-18 shows the basic connections of a standard three-conductor receptacle.

Note the hot wire is connected to the "short slot" in the receptacle and the neutral wire is connected to the "long slot." That too is an NEC requirement. The grounding wire is connected to the round holes in the receptacle, the screw hole that holds

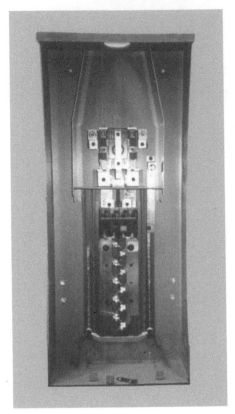

Figure 6-15   Breaker cover removed.

down the cover plate and the bracket that mounts the receptacle to the junction box. Therefore, the cover plate screw is a direct connection to ground. *This screw connection is very important for safely grounding devices that use adapters for connecting older style plugs.*

Ground Fault Circuit Interrupter Receptacles

Figure 6-19 shows the basic connections standard for a *ground fault circuit interrupter* (GFCI) receptacle.

The purpose of the GFCI is to interrupt current flow, should the amount of current flowing out on the hot leg (black) not match the current returning on the neutral (white). The difference only has to be in the order of 5 mA (0.005 A) to trip the breaker.

The GFCI is an essential safety device. The NEC requires GFCI protection to be provided in all bathrooms, kitchens (receptacles within 3 feet of the sink), outdoor, and garage receptacles. Most GFCI receptacles offer an extra set of screws for load connections to additional receptacles to be protected by the same GFCI.

The NEC was changed in 2000 to require that, after 2003, all "bedroom" circuits in a residential installation be served from an *arc fault interrupter* (AFI) type

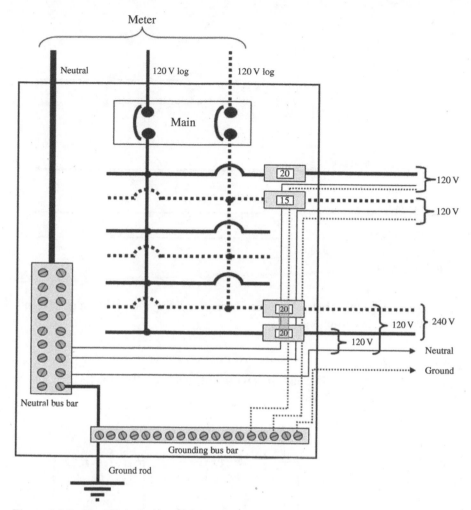

Figure 6-16   Electrical panel–residential.

circuit breaker located in the service panel. This requirement came about because of the concern over the number of fires caused by electric blankets and other warming devices that have deteriorated insulation and arcing thermostats. Most of the time the deteriorated insulation and high activity thermostats allowed arcing and sparking to take place, but there was not enough current to trip a standard circuit breaker. In other words, some thermostats arced too often when deciding whether heating elements should be on or off, thus leading to fires. The AFI detects what is referred to as *electrical noise* generated by arcing and sparking, and thus trips open the circuit breaker. (Note, to avoid arguments, it is normally assumed that if a room has a closet, or is capable of having a closet, it should be considered a bedroom for code compliance, and the outlets in the room should be served through AFI devices.) Since then the NEC has been changed to require an AFCI on basically all 15- and 20-ampere branch

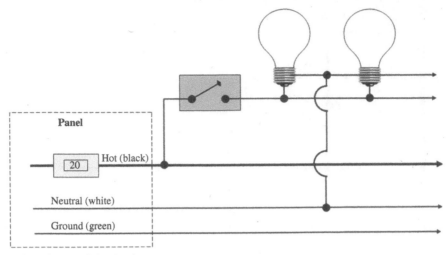

Figure 6-17   Light circuit.

circuits supplying outlets in kitchens, family rooms, dining rooms, living rooms, and recreation rooms in a dwelling.

240 Volt Loads

Figure 6-20 shows the basic connection wiring of standard 240 V loads (i.e., clothes dryers, stoves, and water heaters).

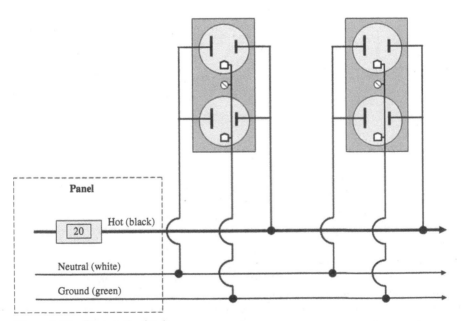

Figure 6-18   Receptacle circuit.

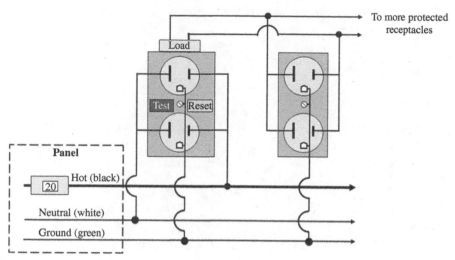

Figure 6-19   GFCI circuit.

Note the two 120 V breakers are bridged together with a plastic cap so that both breakers trip if either one trips. In many cases, a single molded case is used to house the 240 V breaker mechanism. In this case, there are two circuit breakers inside the case, but only one control switch handle.

The neutral (white) wire is brought into the 240 V appliance to be used for any 120 V loads such as lights, clocks, timers. The ground wire (green) is connected to the metal appliance. The green ground wire will cause the 240 V panel breaker to trip, should either hot wire fray or short circuit to the metal appliance.

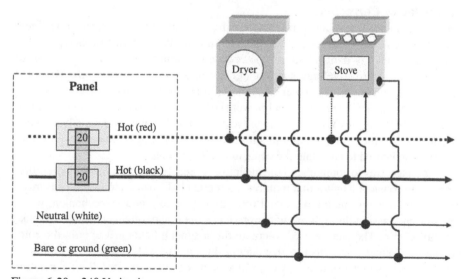

Figure 6-20   240 V circuit.

Figure 6-21   Industrial panel. Reproduced with permission of Photovault.

Commercial and Industrial Service Entrance Equipment

Commercial and industrial service entrance equipment like that shown in Figure 6-21 normally consists of metering equipment (including CTs), a main circuit breaker, disconnect switch, several feeder breakers and sometimes power factor correction capacitors, emergency generator with transfer switch, and an uninterruptible power supply (UPS) system. Some large industrial operations have very large motors requiring soft start (sometimes called "*reduced voltage start*") equipment to reduce the high inrush current to motors when starting.

Power Factor Correction

Low power factor loads such as motors, transformers, and some electronic non-linear loads require reactive energy or "VARs" from the utility to operate properly. Excessive reactive energy demand should be reduced or minimized with installation of capacitors to improve voltage support, reduce losses, lower power bills (in some cases), and improve overall power efficiency (on both sides of the meter).

Power factor is the ratio of real power (i.e., Watts) to the total power (i.e., magnitude of the Watts plus VARs). The reactive portion of the total power can be reduced or eliminated with the application of shunt capacitors. The consumer's power factor information is used to calculate the capacitor requirements.

An easy way to comprehend the meaning of "reactive" power or energy requirements is to consider a motor that requires magnetic fields to operate. A motor is made up of coils of wires and a large metal rotor that spins to produce mechanical work. The current passing through the wires produces the magnet fields required to make the motor spin. The energy used to create the magnetic fields just to spin the rotor is "reactive" energy and this reactive energy does not provide useful work by itself. The "real" component of total power produces the useful work. Installing capacitors to counteract the motor's need for reactive power reduces, minimizes, or eliminates the reactive component of total power from the energy source. The installation of shunt capacitors can help supply the reactive requirements of the motor (i.e., inductive loads) instead of it coming from the generators. The improved power factor from the installation of shunt capacitors at the load center is measured by the power factor metering equipment.

To correct low power factors, the customer and/or the utility install the capacitors. When the utility installs the capacitors, the consumer pays a utility reactive energy fee because the reactive power still flows through the meter. When the consumer installs capacitors on their side of the meter, they no longer pay the extra utility fees. Note, not all utilities charge for low power factors.

Overcorrecting Power Factor with Capacitors

Overcorrecting power factors with excessive capacitance increases the current flowing on the lines above minimum. Similar to when inductive load causes the total current to increase above minimum, capacitors that overcorrect power factor can also increase the total current flow above minimum. When the consumer overcorrects, the extra reactive power flows through the metering equipment, out into the distribution network, and into adjacent consumers' inductive loads; thus reducing the utility's need to install capacitors.

In some cases, the consumer supplying the extra capacitance receives a credit from their utility for excessive reactive power going into the utility system. This extra reactive power is actually used by adjacent consumers. The utility does not have to install as many system capacitors. As a result, the consumer might get a credit on their power bill. Note too much overcorrection can cause high-voltage conditions and power quality issues which can weaken insulation, shorten equipment life, and cause other system problems.

Capacitor banks can be switched on or off based on load requirements, time of day, voltage level, or other appropriate condition to match the reactive power demand of the load. The application of switched capacitors further improves power system and load performance.

Location of Power Factor Correction Capacitors

Typically, utilities do not differentiate between locating power factor correction capacitors on the demand side of the meter. However, the closer the capacitors are installed to the load itself, the more beneficial the results are for the consumer. For instance, the capacitor bank can be located on the demand side of the meter and have the reactive metering register good power factor to the utility or the capacitors can be located adjacent to the load connection terminals reducing the current flow in the consumer's electrical system. The utility is typically only interested in the customer's power factor at the meter. The customer benefits by putting the capacitors as close to the load as possible to minimize losses in the building wiring system and improve terminal voltage so that the equipment can run more efficiently.

Motor Starting Techniques (Optional Supplementary Reading)

When large motors are started, noticeable *voltage dips* or *flicker* can occur on the consumers wiring system that can also effect neighboring loads and the utility's system.

Depending on the voltage sensitivity of other connected loads, these voltage dips can be unnoticeable, annoying, or harmful to the equipment. For example, light bulbs can dim and be annoying to office personnel; however, voltage dips can cause other motor loads to slow down, overheat, and possibly fail. Reduced motor starting equipment is often used to minimize voltage dips and flicker.

The iron core and copper wires in large motors need to become magnetized before running at full speed. The in-rush current required to start large motors create the magnetic fields requiring starting currents to be as high as 7–11 times the full-load current of the motor. Therefore, when large motors start, they often cause low-voltage conditions on the conductors from high current flows. Utilities normally adopt guidelines or policies for starting large motors. When starting a motor exceeds the utility requirement for *voltage dip* or *flicker* (usually set around 3–7%), then special motor starting techniques are usually required.

There are several methods for reducing voltage dip and flicker from large motors starting. Reduced voltage motor starting equipment (i.e., *soft starting*) such as capacitors, transformers, special winding connections, and other control devices are commonly used in motor circuitry to reduce the in-rush current requirements of large motors during start-up conditions.

The three most common means of providing soft starting or reduced voltage starters on large motors are the following:

1. *Resistance* is temporarily placed in series with the motor starter breaker contacts or contactor to cause reduced current to flow into the motor initially when started, then the resistors are shorted out for full running voltage. This approach can reduce the in-rush current to less than five times full-load current. Once the motor comes up to full speed, the resistors are shorted out leaving solid conductors serving the motor power requirements.

2. *Wye-delta* connection change-over in the motor windings is another very effective way to reduce inrush current. The motor windings are first connected wye, where the applied voltage is only line-to-ground, then the motor windings are connected delta for full voltage and output power.

3. *Auto-transformers* are sometimes used to apply a reduced voltage to the terminals when started and then switched out to full voltage after the motor reaches full speed. This scheme can be used with motors that do not have external access to the internal windings.

Emergency Stand-by Generators

Emergency power transfer systems are commonly used to provide local emergency power upon loss of utility power. Upon loss of utility power, the generator, like that shown in Figure 6-22 is immediately started and allowed to come up to speed and warm up before the transfer switch connects the load. Potential transformers (PTs) are used in the transfer switch to sense when the utility power is on and off. These time delays are usually short, approximately 15 seconds to load pick-up after the loss of utility power.

Some consumer emergency generators are used by the utility for online peaking generation. These generators parallel the utility power system and incorporate

Figure 6-22   Emergency generator. Reproduced with permission of Photovault.

special protective relaying schemes to synchronize with the utility. Synchronization requires a proper match among frequency, voltage, phase angle, and rotation before the consumer's emergency generator can be connected to the utility power system.

UPS Systems

Uninterruptible power supply (UPS) systems are typically found in facilities such as police stations, hospitals, control centers, where 100% reliable power is needed. Figure 6-23 *shows* a block diagram of a typical emergency power generator system with a UPS. Notice that utility power feeds all load panels including the Main, Emergency, and UPS panels. Upon loss of utility power, the generator starts immediately. Once the generator is up to speed and able to carry load, the transfer switch operates and connects the generator to the emergency and UPS load panels. Note, the UPS panel loads never experience an outage because those loads are fed by batteries through a dc to ac inverter. The generator begins charging the batteries once the transfer switch operates.

When utility power is restored, all loads including emergency loads are turned off while the transfer switch reconnects utility power to the Main breaker panel. If the transfer scheme includes synchronization provisions, there might not be a need to de-energize the Main breaker panel during the transfer back. When utility power is restored, critical UPS loads again remain powered by the batteries during the transition. The battery charger is reconnected to utility power.

Power Quality

Power quality is the term used to describe fitness of the electric power service to run electrical load and the load's ability to function properly. Without satisfying a minimum level of proper power, an electrical device (or load) may malfunction, fail

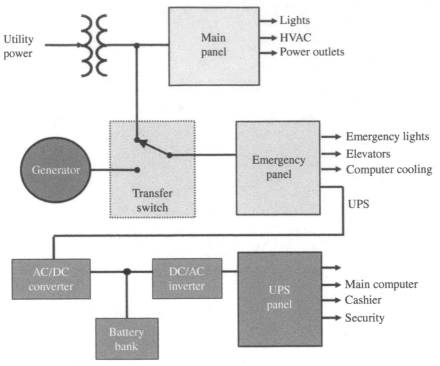

Figure 6-23   UPS system.

prematurely or not operate at all. There are many ways in which electric power can be of poor quality. Note that some loads are not sensitive to power quality issues while others are. Incandescent light bulbs are not affected by minor power quality issues such as voltage spikes, harmonics, while other load types are very sensitive to power quality issues. For example, sophisticated hospital equipment and manufacturing equipment can be vulnerable to power quality issues.

This section provides a high-level overview of the meaning of power quality and service-related issues. There are basically seven types of power quality issues described by IEEE standard 1159, as listed below and shown in Figure 6-24:

1. Transients
   a. Impulsive (surges)
   b. Oscillatory)
2. Interruptions
   a. Instantaneous (0.5–30 cycles)
   b. Momentary (30 cycles–2 seconds)
   c. Temporary (2 seconds to 2 minutes)
   d. Sustained (greater than 2 minutes)
3. Sag/Undervoltage

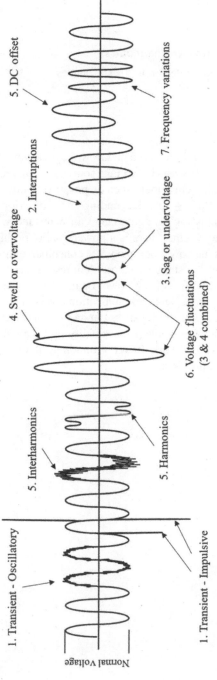

Figure 6-24   Power quality waveforms.

4. Swell/Overvoltage
5. Waveform distortion
   a. DC offset
   b. Harmonics
   c. Interharmonics (induced harmonics)
   d. Notching (periodic interference disturbances)
   e. Noise
6. Voltage fluctuations
7. Frequency variations

Most technological equipment run on low-voltage dc power supplies, which are converted from ordinary 120 V service voltage sources. There are acceptable limits as to the level of power quality equipment such dc power supplies can tolerate. Most dc power supplies tolerate 120 V ± 10%. Depending on the duration of the power quality waveform issue, this technological equipment can withstand wider variations. One would have to determine whether there is a power quality problem and whether the source of the problem is the power company, the consumer, or an adjacent consumer. Each of which would need to share in the problem resolution.

Some of the adverse effects from power quality issues are overheating motors, premature equipment damage, data communication errors, analog interference, false switching operations, and other annoyances depending on the equipment and issues.

The equipment used to resolve power quality issues are uninterruptible power supplies, filters, surge suppressors, better grounding, shielding, and other means depending on the type of problem.

CHAPTER 7

SYSTEM PROTECTION

CHAPTER OBJECTIVES

After completing this chapter, the reader will be able to:

- ☑ *Explain how transmission and distribution lines, substation equipment, and generators are protected*
- ☑ *Discuss what is meant by zones of protection*
- ☑ *Explain the concept of inverse current versus time with respect to clearing faults*
- ☑ *Explain how to effectively use one-line diagrams*
- ☑ *Describe the steps needed to synchronize a generator onto the power grid*

TWO TYPES OF PROTECTION

There are two types of "protection" referred to in electric power systems. The first is *system protection* having to do with protective relays, fault currents, effective grounding, circuit breakers, fuses, etc. The second is *personal protection* having to do with rubber gloves, insulating blankets, grounding jumpers, switching platforms, tagging, etc. This chapter discusses the first one "system protection."

The protection of power system equipment is accomplished by protective relaying equipment that is used to trip circuit breakers, reclosers, motorized disconnect switches, and self-contained protection devices. The objective of system protection is to remove faulted equipment from the energized power system before it further damages other equipment or becomes harmful to the public or employees. It is important to understand that system protection is for the protection of equipment—it is not intended for the protection of people.

System protection protects power system equipment from damage due to power faults and/or lightning strikes. System protection uses solid state and/or electromechanical protective relays to monitor the power system's electrical characteristics and trip circuit breakers under abnormal conditions. Also, the protective relays initiate alarms to system control, notifying operators of changes that have occurred in the system. The control operators react to these incoming alarms from the system protection equipment.

Electric Power System Basics for the Nonelectrical Professional, Second Edition. Steven W. Blume.
© 2017 by The Institute of Electrical and Electronics Engineers, Inc. Published 2017 by John Wiley & Sons, Inc.

Another means for providing equipment protection is proper grounding. Effective or proper grounding can minimize damage to equipment, cause protective relays to operate faster (i.e., open circuit breakers faster), and provides additional safety to personnel.

The explanation of system protection will start by first explaining the different types of protective relays and then proceed to explaining how distribution lines, transmission lines, substations, and generators are protected.

Before we dive into protective relays, let us review the various types of faults these relays are trying to sense. The common fault types are described below:

- *Line-Ground*: the most common fault type is the line-to-ground fault. That fault condition refers to events, such as trees coming in contact with one phase, and bird contact. Regarding underground, these are typically cable, splice, or elbow failures.

- *Phase-Phase*: a less common fault type is the phase-to-phase fault. That condition occurs when wind causes two-phase conductors to come in contact with each other, creating a short circuit and thus tripping the circuit breaker. Phase-to-phase faults can be caused by several conditions, such as ice unloading and car-pole accidents.

- *Three-Phase*: the rarer fault type is the three-phase fault. This condition occurs when all three phases come in contact with each other to trip the circuit breaker. Three-phase faults can occur with airplane accidents, equipment failures, and others.

- *Open-Conductor*: when a three-phase line or equipment experiences a conductor failure in the open condition and the other two conductors remain intact, unbalanced currents flow in the system that must be sensed and the condition removed. Large motors, for example, can be damaged, should one of the three phases become de-energized; this is known as *"single phasing."*

- *Others*: there are several other events that can occur on power systems that trip circuit breakers. Faults are usually referred to equipment failures, power lines that fail due to environmental reasons, or facility failures. Storms can also initiate faults on electrical systems where the circuit breakers and fuses must isolate the failed equipment.

SYSTEM PROTECTION EQUIPMENT AND CONCEPTS

System protection, often called *protective relaying* is composed of relay devices in substations that monitor the power system's voltages and currents through the CTs and PTs and are programmed to initiate "trip" or "close" signals to circuit breakers if the pre-programmed thresholds are exceeded. System control operators are then alarmed of the new state or conditions. The relays, trip signals, circuit breaker control systems, and the system control equipment are all battery powered. Therefore, the entire system protection operation is functional, should the main ac power system be out of service or become out of service.

Protective Relays

A protective relay is a device that monitors system conditions (amps, volts, Watts, etc. using CTs and PTs) and reacts to the detection of abnormal conditions. The relay compares the real-time actual quantities against preset programmable threshold values and sends dc electrical control signals to trip circuit breakers or other opening devices in an effort to clear an abnormal condition on the equipment it is protecting. When system problems are detected and breakers are tripped, alarm indications are sent to system control operators and sometimes other protection operations are initiated. As a result, equipment might be de-energized and consumers out of power with minimal equipment damage. The operation of protective relays is the stabilizing force against the unwanted de-stabilizing forces that occur in electric power systems when something happens, such as unanticipated power faults and lightning strikes.

Protective relays are manufactured as two types; *electromechanical* and *solid state*. The older type electromechanical relays are composed of coils of wire, magnets, spinning disks, and moving electrical switch contacts and are very mechanical in nature. The newer type solid-state relays are electronic and have no moving parts. Solid-state relay systems are often referred to as microprocessor type relays since they can accomplish many functions in one box. These microprocessor type relays are the most up to date relay type. These solid-state relays have revolutionized protection and their untapped capability has stretched over to substation automation, digital integration, and Smart grid functionality; not to mention the tremendous saving in space requirements in substation control buildings. Most utilities have replaced their electromechanical relays with microprocessor-based solid-state relays.

The basic differences are listed below:

Solid State

- Advantages: multiple functionality, small space requirements, easy to set up and test, self-testing, remote access capability, and they provide fault location information. See Figure 7-1.
- Disadvantages: external power required, software can be complex and many "functional eggs" all in one basket.

Electromechanical Relays

- Advantages: usually self-powered, simple, and single function design. See Figure 7-2.
- Disadvantages: normally one relay per phase, difficult to set up and adjust, and requires frequent testing.

Inverse Current-Time Concept

Typically, protective relays are designed to follow the *inverse current-time* curve as shown in Figure 7-3. In other words, *the time to trip a circuit breaker shortens as the amount of fault current increases*. Therefore, a relay sensing a fault located near a substation would initiate a trip to the breaker faster than if the fault were located

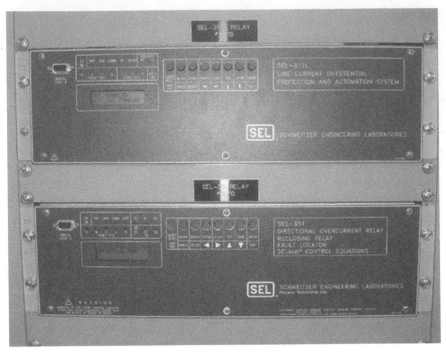

Figure 7-1   Solid-state relays.

down the line because less fault current flows due to the additional resistance of the wire. Note, each circuit breaker has a fixed amount of time to open a circuit once it receives a trip signal from the relay. Some breakers trip in less than 2 cycles after receiving a trip signal while some older breakers might take up to 9 cycles to trip.

The time to trip is shown along the horizontal axis and the amount of current flowing in the line (e.g., CT) is shown along the vertical-axis. When the actual real-time current is below the horizontal set point portion of the curve or the *minimum pickup setting*, the time to trip becomes *never* and the relay does not operate. When the current exceeds the *instantaneous setting* on the curve, the time to trip becomes *as fast as possible* and the relay initiates a trip command to the breaker without any intentional time delay. Between these two set points, the relay engineer adjusts the shape of the curve to meet various *system protection coordination* objectives.

Relay coordination is the term used to create a situation where the most *downstream* clearing device from the source clears the fault first, thus minimizing the number of customers out of service. Whenever possible, the *upstream* devices act as *backup clearing* devices. The proper coordination of all the protective relay devices in the transmission and distribution systems, or even a single power line is a very special art and the relay engineer must understand all the idiosyncrasies associated with all the equipment and conditions possible. There are many key factors that play very important roles in the proper design and coordination of protective relaying.

Figure 7-2   Electromechanical relays.

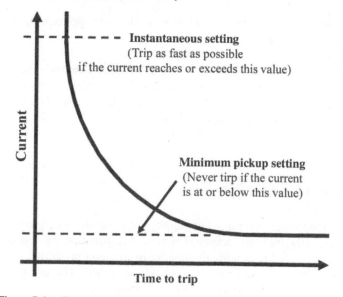

Figure 7-3   Time versus. current curve.

One-Line Diagrams

A *one-line diagram* (also referred to as the *single-line diagram* "*See* One-line diagram") is a simplified drawing of the system or a portion of the system that shows the electrical placement of all major equipment. One-line diagrams are actually simplified *three-line diagrams* "*See* One-line diagram" with drawing redundancy removed. Extra information is added to the one-line diagram to give the engineer or systems operator the full picture of the electrical system, including the system protective relay schemes. One-line diagrams are very useful for planning maintenance activities, rerouting power after a fault, switching orders to change system configurations, and to view the relationships between smaller sections of the power system to the overall system.

There are many uses of one-line diagrams. Electric utility personnel use one-line diagrams to perform their work activities on a daily basis. Some of the most common uses are discussed below:

➤ Line crews refer to one-line diagrams to know what protection devices are used on the power line being worked, to identifying disconnect switch locations for load transfer operations and to see the relationship to other nearby lines or equipment that are part of the system in question.

➤ System operators use one-line diagrams to identify the electrical placement of breakers, air switches, transformers, regulators, etc. in substations that may indicate alarms and/or needs corrective action. They use one-line diagrams to figure out how to switch the system equipment to sectionalize failed equipment and to restore power.

➤ Electrical Engineers use one-line diagrams to understand system behavior and to plan for or make changes to the power system to improved reliability and performance.

➤ Consumers use one-line diagrams to identify their electrical equipment, circuits, and protection apparatus.

An example of a simple one-line diagram for a distribution substation is shown in Figure 7-4. Note the protective relay numbers in circles. These numbers represent relay functions and are identified in the adjacent table. A complete list of relay number identifications is available through the IEEE (IEEE Standard C37.2 Standard for Electrical Power System Device Function Numbers).

DISTRIBUTION PROTECTION

Distribution lines (i.e., feeders) are normally fed radially out of substations, meaning distribution feeders have only one utility source. The typical distribution line protection schemes used on radially fed lines normally involve overcurrent protection with reclosing relays and in several cases underfrequency load shed relays. This approach to distribution protection is very common; however, variations do exist.

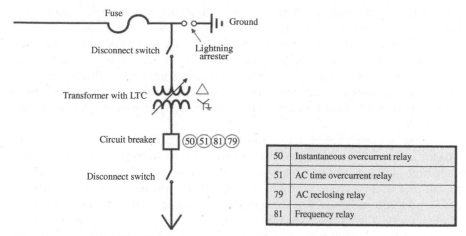

Figure 7-4   One-line diagram.

Overcurrent and Reclosing Relays

Each distribution feeder has a set of overcurrent relays; one for each phase and one for ground overcurrent for a total of four overcurrent relays. Each relay has an instantaneous and a time delayed capability. The instantaneous and time delay capabilities are interconnected with the reclosing relay. This typical substation relay package must also coordinate with the downstream reclosers and fuses located on the feeder itself.

The overcurrent relays are connected directly to current transformers (i.e., CTs) located on the circuit breaker bushings. This enables the monitoring of actual current magnitudes flowing through the breaker in real time. Normally there are four CTs used for each feeder breaker (one for each phase and one for the grounded neutral). Each overcurrent relay has both an instantaneous and a time delay overcurrent relays connected to the CTs. These relays are looking for feeder faults that are phase–to–ground, phase-to-phase, two phases to ground, or three phases together. The protection engineer analyzes the available fault current magnitudes for each feeder breaker and recommends relay settings that are later programmed into the relays. These relay settings are periodically evaluated and tested to make sure they operate properly.

Typical Distribution Relay Operation

Suppose a lightning strike hits a distribution feeder's "B" phase near the substation and causes a B-phase to ground fault. The lightning arrester would fire and cause a high line-to-ground fault. The ground overcurrent relay would sense the increase in ground current through the transformer's neutral CT and instantaneously send a trip signal to that feeder's breaker. The breaker trips the line and all consumers on the line are now out of power. The overcurrent relay simultaneously sends a signal to the reclosing relay to initiate a timer. After the preset time delay expires in the reclosing relay timer, the reclosing relay contacts close and initiates a close signal to the same breaker, thus re-energizing the feeder. This first time delay, or interval, is typically

5 seconds long. If the fault is temporary, as in lightning, all consumers will remain back in service after the brief outage.

A comment about the above scenario; the instantaneous trip setting (sometimes referred to as the *fast trip* setting) on the substation breaker is faster than the time it takes to melt a downstream fuse. When a lightning strike hits a distribution line, the normal sequence of events would be to have all consumers trip offline and about 5 seconds later all consumers are back in service without having any distribution fuses melt.

Now suppose a tree got into a distribution feeder lateral downstream of a fuse. The feeder breaker would trip on instantaneous or fast trip and recloses about 5 seconds later. However, this time the tree is still in the line and the short circuit current flows again after the reclose because of the tree remaining in the line. In most distribution protection schemes, the instantaneous trip setting is taken out of service after the first trip and the time-delayed overcurrent relay takes over. The removal of the instantaneous or fast trip relay after the first trip and reclose allows time for fault current to melt the fuse and clear the fault on the fused lateral only. Therefore, only those consumers downstream of the blown fuse are out of power. All the other consumers on the feeder experience a voltage sag during the time the fuse is melting and then full voltage is resumed after the fuse melts. The customers downstream of the fuse remain out of power until a line worker from the power company finds the blown fuse, clears the tree, replaces the fuse, and closes the cut-out with the new fuse to restore power.

In the cases where the fault (i.e., tree) is on the main feeder, not on a fused lateral, the substation breaker will trip by the instantaneous relay. After the first time delay of about 5 seconds, the reclose relay sends a close command to the substation breaker to re-energize the feeder. If the tree is still in the line after the reclose, the breaker will trip again by the time delay overcurrent relay. After another preset time delay (about 15 seconds) the reclose relay sends another close command to the breaker to re-energize the feeder again. By this time, the tree branch may have been cleared from the fault. If not, the fault current flows again and the time delay overcurrent relay trips the feeder for the third time. All consumers are out of power again. Now after another time delay (this time maybe 25 seconds) the line is automatically reclosed for the fourth time. If the fault is still present, the overcurrent relay trips the breaker for the fourth time and *locks-out*. The reclosing relay no longer sends a close signal to the breaker and all customers remain out of power until the line workers clear the fault (i.e., the tree), reset the relays, and close the breaker. (Note: there is a programmable reset timer that places the sequence back to the beginning, that is, initial trip being fast, if the fault were temporary.)

As stated earlier, there are variations to this distribution scheme; however, what was described above is very common in the industry. *Caution*: as just described, a distribution line can become re-energized several times automatically. A similar scenario would occur in a car-pole accident where a power line conductor comes in contact with the car. The conductor could and probably will re-energize multiple times before the breaker "locks-out." Also, system control operators could test the line remotely to see if the line will remain energized. After a lock-out condition, a line worker is sent out to *patrol the line* only to discover that the problem is a car-pole accident. The example above illustrates why *it is very important to realize that a power line can be re-energized automatically at any time and to always stay clear of a fallen line power!*

Underfrequency Relays

In an effort to stop or prevent a cascading outage, *underfrequency relays* are used to shed load when the system frequency is dropping. Underfrequency relays are also referred to as *load shed* relays. The system frequency will drop if there is more load than there is generation (i.e., load-generation unbalance). When generation or an important tie line is tripped, system frequency can drop and load shed relays will start to trip feeder breakers as a *remedial action* to balance load and generation. This automatic load shedding scheme can trip up to 30% of total load, in steps, in an effort to prevent the system from cascading into a wide area outage.

Keep in mind that the standard frequency in the United States is 60 Hz, the typical underfrequency relay settings are chosen based on the following guidelines:

At 59.3 Hz, shed a minimum of 10% of load.

At 59.0 Hz, shed a minimum of 10% of load.

At 58.7 Hz, shed a minimum of 10% of load.

At 58.5 Hz or lower, the system may take any action it deems necessary, including a domino effect disturbance.

Some systems start diesel engine generators and/or combustion turbines automatically upon underfrequency detection in preparation to come online to help restore the generation-load balance. All of these remedial action schemes are intended to balance generation and load and stop the possibility of a cascading outage disturbance.

TRANSMISSION PROTECTION

Transmission protection is much different than distribution protection simply because transmission is usually not radially fed. Normally transmission systems have multiple feeds to a substation and transmission lines must have special protective relaying schemes to identify the actual faulted transmission line. To complicate matters, some transmission lines might have generation at the other end that contributes to the fault current while others are transporting generation from different lines and substations. Further, some transmission lines are only serving load at their far end. The application or concept of *zone relaying* (sometimes called *distance or impedance relaying*) with directional overcurrent capability is used to identify and trip the faulted transmission line.

The direction of the fault current verifies that a particular breaker(s) need to trip. For example, the excessive current must be leaving the substation toward the fault, opposed to just excessive current magnitude. Both fault current magnitude and direction are required for transmission breakers to trip.

As another example, notice the location of the fault on the transmission one-line diagram in Figure 7-5. Notice the multiple transmission lines, generators,

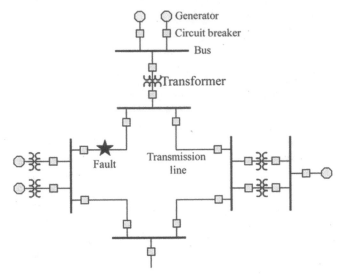

Figure 7-5   Transmission fault.

transformers, and buses for the power system. A fault on one of the transmission lines requires breakers on both ends of that line to trip. Zone relaying identifies the faulted line and trips the appropriate breakers. Also, zone relaying provides backup tripping protection, should the primary protection scheme fail.

Zone or Distance Relays

Figure 7-6 shows the concept of zone relaying.

In this particular scheme, each breaker has three protection zones. For example, breaker "A" has three zones looking toward the right (as shown), breaker "B" would

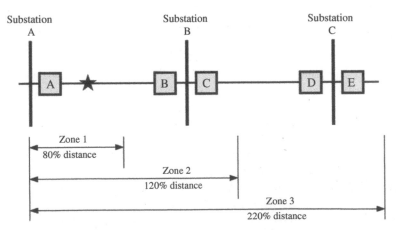

Figure 7-6   Zone protection.

have three zones looking left (not shown), breaker "C" would have 3 zones looking right, and so on. Typical zone relay settings are as follows:

Zone 1 Relays

The Zone 1 relay is programmed to recognize faults that are located 80–90% of the line section and trip instantaneously (i.e., 1–3 cycles).

In this example, the fault is in Zone 1 of breaker "A" and therefore breaker "A" is tripped at high speed. High speed implies that the relay is set for instantaneous tripping of the breaker and fault clearing depends only on the time it takes for the breaker to open its contacts and interrupt the current.

Zone 2 Relays

The Zone 2 relay is programmed to recognize faults that are located about one-line section plus about half of the next line section (approximately 120–150%). The trip is time delayed to coordinate with Zone 1 relays.

In this example, the fault is in Zone 2 of breaker "B" and would trip after a short time delay. However, in zone protection schemes, fiber optic, microwave, power line carrier, or copper circuit communications systems are used to send a signal to the line's opposite end breaker, when fault current is flowing in that direction. In this case, breaker "A" would send a transfer trip signal to substation B, telling breaker "B" to bypass its Zone 2 time delay setting and trip immediately. This provides high speed line clearing from both ends even though there is a built in time delay in Zone 2 relays.

Note, if the fault were in the middle of the line, both ends would trip Zone 1 at high speed.

Zone 3 Relays

Zone 3 relays are set to reach the protected line section plus the next line section plus an additional half line section as a backup (approximately 250%). The trip is time delayed more than Zone 2 to coordinate with Zone 2 and Zone 1 protection. Zone 3 provides full backup.

In the example above, Zone 3 backup protection would not be involved. Should a Zone 2 breaker fail to trip the line, then Zone 3 would trip as a backup.

Note, the various types of telecommunication systems used in electric power systems for system protection schemes like this are discussed in Chapter 9.

SUBSTATION PROTECTION

Substation protection is generally accomplished using *differential relays*. Differential relays are used to protect major transformers and bus sections from faults. Substation differential relays are very similar in concept to GFCI breakers discussed earlier in residential wiring in Chapter 6. In the case of the GFCI receptacle breaker, the current leaving the hot leg (black wire) must equal or be within 5 mA of the current returned in the neutral (white wire) or the GFCI breaker will trip. Similarly, differential relays used in substation transformers and buswork monitor the current entering versus exiting the protection zone. These concepts are discussed below as they apply to substation transformers and bus protection schemes.

Differential Relays

Differential relays are generally used to protect bus, transformers, and generators. Differential relays operate on the principle that the current going into the protected device must be equal to the current leaving the device or a differential condition is present. Should a differential condition be detected, then all source breakers that can feed fault current on either side of the device are tripped.

Transformer Differential Relays

Current transformers (CTs) on both the high side and low side of the transformer are connected to a *transformer differential relay*. *Matching CTs* are used to compensate for the transformer windings turns ratio. Should a differential be detected between the current entering the transformer and exiting the transformer after adjusting for small differences due to losses and magnetization, the relay trips the source breaker(s) and the transformer is de-energized immediately. There is no automatic reclosing on transformers (or bus).

Bus Protection Schemes

Bus differential relays are used to protect the bus in a substation. The current entering the bus (usually exiting the power transformer) must equal the current leaving the bus (usually the summation of all the transmission or distribution lines). Line-to-ground faults in the bus will upset the current balance in the differential relay and cause the relay contacts to close, thus initiating trip signals to all source breakers.

Over and Under Voltage Relays

Another application of system protective relays is the monitoring of high and low bus voltage. For example, *over voltage relays* are sometimes used to control (i.e., turn off) substation capacitor banks. While, *under voltage relays* are sometimes used to switch on substation capacitor banks. Over and under voltage relays are also used to trip line or generator breakers due to other abnormal conditions.

GENERATOR PROTECTION

The chances of failure of rotating machines are small due to improved design, technology, and materials. However, failures can occur, and the consequences can be severe. It is very important that proper generation protection is provided. This section summarizes the techniques used to protect very expensive generators.

When a generator trips offline for any reason, it is extremely important to determine exactly what caused the generator to trip. This condition should not happen. Some of the undesirable operating conditions for a generator to experience and the protective scheme or device used to protect the generator are listed below:

Winding Short Circuit

Differential relays normally provide adequate protection to guard against shorted winding in the generator stator. The current entering the winding must equal the current leaving the winding or a winding-to-ground fault may be present and the generator breaker is tripped.

Unbalanced Fault Current

The very strong magnetic forces that are imposed on a generator during fault conditions, especially an *unbalanced fault* (e.g., line-to-ground fault as opposed to a three-phase fault), cannot be sustained for a long period of time. This condition quickly causes rotor overheating and serious damage. To protect against this condition, a reverse rotation overcurrent relay is used to detect these conditions. Reverse rotation (i.e., *negative sequence*) relays look for currents that want to reverse the direction of the rotor. *Positive sequence* currents, for comparison, rotate the rotor in the correct direction.

Frequency Excursion

A generator's frequency can be affected by over and under loading conditions and by system disturbances. Frequency excursions cause possible over excitation problems and can cause turbine blade damage. Excessive underfrequency excursion conditions can also affect auxiliary equipment such as station service transformers that power ancillary equipment at the power plant. Underfrequency relays and Volts per Hertz relays are often used to protect against excessive frequency excursions.

Loss of Excitation

When loss of generator excitation occurs, reactive power flows from the system into the generator. Complete loss of excitation can cause the generator to lose synchronism. Therefore, loss of excitation (i.e., under voltage relay) is used to trip the generator.

Field Ground Protection

Field ground protection is needed to protect the generator against a possible short circuit in the field winding (i.e., a fault between the rotor winding and stator winding). A fault in the field winding could cause a severe *unbalance and generator vibration* that could possibly damage the generator's rotor shaft.

Motoring

This condition is attributed to insufficient mechanical energy onto the shaft by the prime mover. When this occurs, power flows from the system into the generator, turning the generator like a motor. Motoring can cause overheating of the turbine blades. Protection against generators acting in a *motoring condition* is highly desirable and usually results in tripping the generator.

GENERATOR SYNCHRONIZATION

The purpose of a synchronizing relay is to safely connect two 3-phase lines together or to place a spinning generator online. Figure 7-7 shows a generator breaker that needs to be closed. There are *four* conditions that must be met first in order to safely connect two 3-phase systems together gracefully. Failure to meet these four conditions can result in catastrophic failure of the equipment (i.e., generator). Note, *permissive relays* are used in circuits like this to block the closing of circuit breakers until all conditions are met. An analogy to "permissive relays" would be to require that your seat belt be buckled before your car will start.

The four conditions that must be met prior to closing a breaker to synchronize a generator to a running system are summarized below:

Frequency
The generator must have the same frequency as the system before the circuit breaker can be closed. Not matching the frequency on both sides of the breaker before closing could cause the generator to instantly speed up or slow down causing physical damage or excessive power transients.

Voltage
The voltage must be close to the same magnitude on both sides of the breaker connecting the systems together. Widely differing voltages could result in excessive voltage transients.

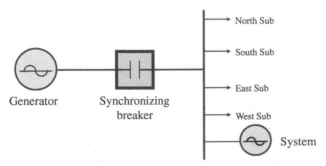

Figure 7-7   Generator synchronization.

Phase Angle

The relative phase angle of the generator must be equal to (or very close to) the phase angle of the system before the synchronizing breaker can be closed. (Note, it is only necessary to match one phase on both sides of the breaker so long as it is the same phase.) To add clarification to this important condition, the frequency on both sides of the breaker could be 60 Hz, however one side might be entering the positive peak of the cycle while the other side is entering the negative peak. This is an unacceptable condition. Both sides must be very close in phase angle (typically less than 15 degrees of the 360-degree cycle) before the breaker can be closed.

Rotation

Rotation is normally established during installation. Rotation has to do with matching phases A, B, and C of the generator with phases A, B, and C of the system. Once the rotation has been established, this situation should never change.

Synchronizing Procedure

Synchronizing relays and/or synchroscopes such as the one shown in Figure 7-8 helps match the generator to the system for a graceful connection. Synchroscopes display the relative speed of the generator with respect to the system. A needle rotating clockwise indicates the generator spinning slightly faster than the system. The normal procedure for closing the breaker is to have the generator spinning slightly faster than the system or at least accelerating in the positive direction when the breaker is closed. Once the breaker is closed, the needle stops spinning. Therefore, the generator is ready to immediately output power into the system.

Figure 7-8   Synchroscope.

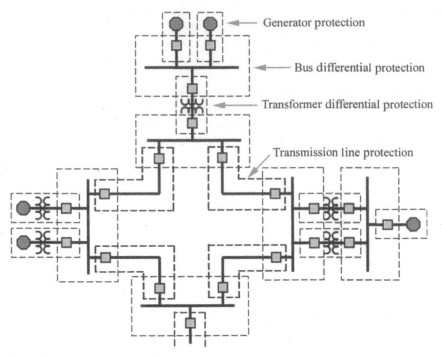

Figure 7-9   Transmission protection.

OVERALL TRANSMISSION PROTECTION

The drawing in Figure 7-9 shows the many zones of protection found in a major inter-connected electric power system. All the *zones overlap* to provide full complement of protection against line, bus, generator, and transformer faults. Overlap is achieved using CTs on opposite sides of equipment being protected.

SUBSTATION AUTOMATION

The advent of Ethernet communications connectivity and smart intelligent devices, including smart protective relays is rapidly changing how substation protection is configured, constructed, maintained, and operated. New equipment being used in substations include digital interface instrument transformers (CTs and PTs), digital interface protective relays, and fiber optic linked equipment. New intelligent circuit breakers provide this capability or existing equipment can be modified with *"merging units."* Merging units (see Figure 7-10) provide an electronic interface between older traditional analog equipment and new digital interface equipment. These units help transition older substations with new technology devices. The new equipment improves

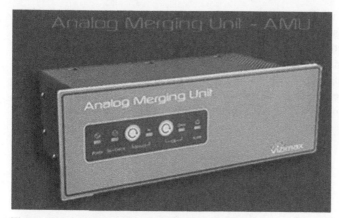

Figure 7-10   **Merging unit.** Courtesy of Visimax.

reliability, connectivity, device programming flexibility, and enterprise information utilization and management.

Digital substation automation helps provide special protection and transfer schemes; such as automatically transfer load from faulted bus to healthy bus while avoiding overload due to adaptive relaying. Intelligent supply sectionalizing schemes using smart devices can identify a faulted section of supply, isolate that section of line, and restore supply to unfaulted sections. New equipment condition monitoring devices watch the number of breaker operations, accumulated fault clearing statistics,

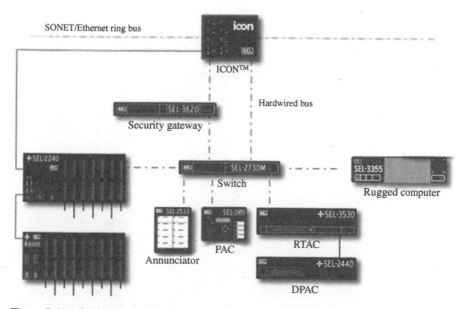

Figure 7-11   Digital substation. Courtesy of Schweitzer Engineering Laboratories, Inc.

etc. to determine when maintenance is required. Effective, just in time maintenance programs are made possible with new information tracking that is provided by modern substation automation devices.

Substation automation and digital connectivity are underway. Figure 7-11 shows the typical arrangement using digitally interfaced equipment in substations using modern LAN Ethernet connectivity over high speed category rated copper wire, fiber optic, and sometimes wireless communications technologies.

CHAPTER **8**

INTERCONNECTED POWER SYSTEMS

CHAPTER OBJECTIVES

After completing this chapter, the reader will be able to:

- ☑ *Explain why interconnected power systems are better than isolated control areas*
- ☑ *Describe the major power grids in North America*
- ☑ *Explain power grid reliability, stability, and voltage control*
- ☑ *Discuss system demand and generator loading*
- ☑ *Explain the purposes of "Spinning Reserve" and "Reactive Supply"*
- ☑ *Discuss what system control operators do to prevent major disturbances*

INTERCONNECTED POWER SYSTEMS

Interconnected power systems (i.e., *power grids*) offer many important advantages over the alternative of independent power islands. Large power grids are built to take advantage of electrical *inertia* for the purpose of maximizing system stability, reliability, and security. (Inertia is discussed later in this chapter.) Also, in today's regulatory atmosphere, large interconnected power grids offer new opportunities in sales/marketing, alternative revenue streams, and resource sharing for a price.

Electric power systems became interconnected power grids a long time ago. Interconnected systems stabilize the grid, which in turn improves reliability and security. It also helps reduce the overall cost of providing reserves. Interconnected systems help maintain frequency, avoid voltage collapse, and reduce the chance of undesirable load shed situations.

Further, interconnected power companies benefit from information exchange opportunities and equipment sharing (including spare equipment). These benefits include joint planning studies, mutual cooperation during emergencies (such as storm damage), and sharing new technologies especially in the areas of telecommunications, system control centers, and energy management.

Electric Power System Basics for the Nonelectrical Professional, Second Edition. Steven W. Blume.
© 2017 by The Institute of Electrical and Electronics Engineers, Inc. Published 2017 by John Wiley & Sons, Inc.

Please note the emphasis of this chapter is on electrical fundamentals of interconnected power system operations; the regulatory and power agency organizations aspect will be addressed but not elaborated.

THE NORTH AMERICAN POWER GRIDS

The *North American Electric Reliability Corporation* (NERC) is responsible for ensuring that the bulk electric power system in North America is reliable, adequate, and secure. NERC was formed in 1968 and has operated successfully as a self-regulatory organization, relying on reciprocity and the mutual self-interest of all those involved in the production, transmission, and distribution of electricity in North America. NERC has recently acquired the duties of overseeing operating standards compliance with enforcement powers given by the Federal Energy Regulatory Commission (FERC).

The massive interconnected power grid system in the United States and Canada is broken down into four separate grids; the western grid, the eastern grid, Quebec, and Texas.

Figure 8-1 shows the power grid interconnections structure in North America.

The three U.S. grids are composed of regions and/or utilities having interconnected transmission lines and control centers. They share similarities, such as 60 Hz

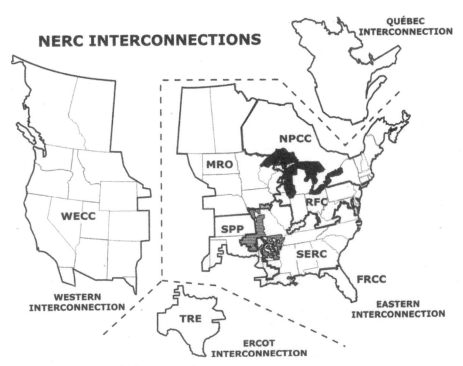

Figure 8-1   Power grid interconnections.

frequency and system transmission voltages, yet they have specific individual requirements such as ownership, topography, and fuel resources. All the generation units in
each grid are synchronized together, sharing total load and are providing reliable generation and load balance within very large power grids.

REGULATORY ENVIRONMENT

The regulatory environment in the electric power industry continues to change, causing some uncertainty in the way companies are structured. Most electric companies
are trying to establish or position themselves as being generation, transmission, or
distribution companies to align with the new regulatory framework.

Due to the governmental changes that have resulted in a deregulated electric
power industry, and to avoid potential conflicts, employees in wholesale power contracts departments must remain physically separated and note communicating from
employees dealing with the actual electrical generation and transmission systems
because of the unfair advantages or disadvantages in an open market environment.
Some view having knowledge of a company's strengths, weaknesses, and future construction projects is unfair. Similar rules exist for the separation of transmission and
distribution of employees where necessary.

Figure 8-2 illustrates where the actual divisions occur in the deregulated model.
Note the division is between the windings of the transformers. However, actual equipment ownership arrangements are defined on a case–by-case basis.

Independent System Operators (ISO) and Regional Transmission Operators (RTO)

The *Federal Energy Regulatory Commission* (FERC) now requires that power entities
form joint transmission operations areas known as *Regional Transmission Operators*
(RTO's) or *Independent System Operators* (ISO's). These groups are charged with
the requirements that all parties work together, have equal access to information, and
provide a marketplace for energy exchange.

In the United States an ISO is a federally regulated regional organization which
coordinates, controls, and monitors the operation of the electrical power system of a

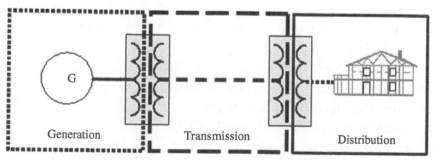

Figure 8-2   Regulatory divisions.

particular service area, typically a single state. The RTOs such as the Pennsylvania—New Jersey—Maryland Interconnection (PJM) have a similar function and responsibility but operate within more than one U.S. state.

The ISO or RTO act as a marketplace in wholesale power now that the electricity market has been deregulated since the late 1990s. Most ISOs and RTOs are set up as non-profit corporations using a governance model developed by the FERC in April 1996. Also, FERC Order 888/889 required *open access* of the grid to all electricity suppliers and mandated the requirement for an *Open Access Same-Time Information System* (*OASIS*) to coordinate transmission suppliers and their customers.

The Canadian equivalent to the ISO and RTO is the *Independent Electricity System Operator* (IESO).

There are currently five ISOs operating in North America:

> Alberta Electric System Operator (AESO)
> California ISO (CAISO)
> Electric Reliability Council of Texas (ERCOT), also a Regional Reliability Council (see later)
> Independent Electricity System Operator (IESO), operates the Ontario Hydro system
> New York ISO (NYISO)

There are currently four RTOs operating in North America:

> Midwest Independent Transmission System Operator (MISO)
> ISO New England (ISONE), an RTO despite the ISO in its name
> PJM Interconnection (PJM)
> Southwest Power Pool (SPP), also a Regional Reliability Council (see later)

Regional Entities

The North American Electric Reliability Corporation (NERC), whose mission is to improve the reliability and security of the bulk power system in North America works with eight regional entities. The members come from all segments of the electric industry: investor-owned utilities; federal power agencies; rural electric cooperatives; state, municipal and provincial utilities; independent power producers; power marketers, and end-use customers. These regional entities account for virtually all the electricity supplied in the United States, Canada, and a portion of Baja California Norte, Mexico.

> Electric Reliability Council of Texas, Inc. (ERCOT)
> Florida Reliability Coordinating Council (FRCC)
> Midwest Reliability Organization (MRO)
> Northeast Power Coordinating Council (NPCC)
> Reliability First Corporation (RFC)
> SERC Reliability Corporation (SERC)

> Southwest Power Pool, Inc. (SPP)
> Texas Reliability Entity (TRE)
> Western Electricity Coordinating Council (WECC)

The Balancing Authority

The NERC rules require that all generation, transmission, and load operating in an interconnection must be included in the metered boundaries of a *balancing authority*. Before deregulation, a balancing authority was almost synonymous with a utility company. The utility company controlled transmission, generation, and distribution, thus was responsible for the balance of all generation and load. By definition, all of the generation, transmission, and load for that utility were inside the control area of the utility, in essence a balancing authority. However, with today's deregulation, balancing authorities are not necessarily individual utility control areas. Balancing authorities are approved by NERC and they may control generation in multiple utilities.

The balancing authority is responsible for maintaining online generation reserves in the event that a generator trips offline. Also, the balancing authority must be capable of controlling generation through the *automatic generation control (AGC)* system. The balancing authority is also responsible for communicating electronically all data required to calculate the *area control error (ACE)*, the difference between scheduled and actual tie-line flow. (Note, AGC and ACE are discussed in more detail later in this chapter.)

Figure 8-3 shows the NERC Regions and balancing authorities.

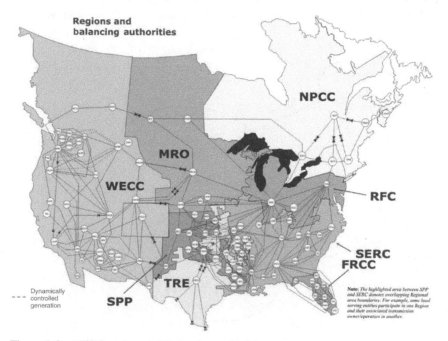

Figure 8-3   NERC regions and balancing authorities.

INTERCHANGE SCHEDULING

In reference to Figure 8-4, the net power flowing on all the *tie lines* between the islands A, B, C, and D in this interconnection must add up to zero. Or, the net interchange of a single company in an interconnected system is equal to the sum of the tie line flows of that company to other companies. The power flowing on these tie lines is accurately metered and scheduled with agreements on pricing. Pricing agreements include provisions for special circumstances, such as emergencies, planned outages, and *inadvertent* power flow. The error between scheduled and actual power flow (i.e., inadvertent) is properly accounted for and settled between the parties involved on a continuous basis.

Area Control Error

The term *ACE* (*area control error*) is used to describe the instantaneous difference between a balancing authority's net actual interchange flow and the scheduled interchange flow, taking into account the effects of frequency and metering error. The term *flat tie line control* is used when only tie line flows are closely monitored in consideration of the actual interchange flow. The term *flat frequency control* is used when only frequency is carefully controlled. When both tie line flow and frequency are carefully controlled by AGC, the term is called *tie line bias*. *Tie line bias* allows the *balancing authority* to maintain its interchange schedule and respond to interconnection frequency errors caused by generation-load unbalance.

The AGC system is controlled by the *Energy Management System* (EMS). The EMS ramps up/down generators to match changing demand taking into consideration

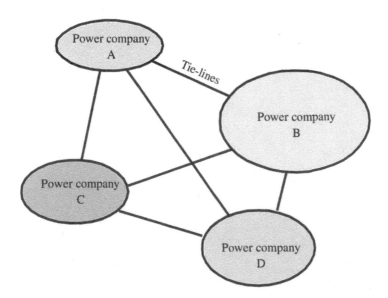

Figure 8-4   Interconnected systems.

several factors, such as cost, efficiency, contingency analysis, and other factors. (Note the computer software tools used by system operators that make up the EMS are discussed in more detail in Chapter 9).

Tie line bias is carefully monitored and reported for all tie lines. Bias is the accepted standard operating constraint for controlling ACE under normal steady-state conditions. Carefully monitoring and adjusting tie line flow helps keep the interconnected system stable. ACE is measured in MW. If the ACE is greater than zero, for example, means the entity is outputting power and if ACE is below zero, the entity is under-generating and importing power. One interesting aspect about the way the ACE equation was derived is that the ACE result remains approximately the same (zero) if a generator trips offline outside the balancing authority area (yet inside the grid interconnection) and goes negative if a generator trips inside the balancing authority area. Therefore, entities know if they need to respond to generation deficiencies (disturbances) by utilizing generation reserves based on the results of the ACE equation immediately following a disturbance.

Time Correction

The power grids adjust their generation pattern to make sure real time (measured in seconds) match grid frequency (i.e., 60 cycles per second). *Time error* is the difference between the time measured at the *balancing authority*(ies) based on 60 cycles per second and the time specified by the National Institute of Standards and Technology. Time error is caused by the accumulation of frequency error over a given period of time. Therefore, adjusting bulk generation, hence shaft speed, for the entire interconnection faster/slower for a specific period of time corrects time error. Ultimately, the number of cycles created by interconnection generation matches the number of cycles that should have been produced over the same period of time (usually daily). For example, if the grid over generated the number of cycles for a given time period, the grid frequency is reduced by 0.02 Hz (59.98 Hz) until corrected. If the grid under generated the number of cycles for a given period, the grid frequency would be increased 0.02 Hz (60.02 Hz) until corrected.

In other words, there are 60 cycles in a second, 3600 cycles in a minute and 5,184,000 cycles in a day. The grid frequency must increase or decrease slightly (±0.02 Hz) if the actual number of cycles generated does not match the exact same number of cycles that should have been generated based on real time. Time correction is a very important concept that must be met on a daily basis. The frequency is closely monitored at key locations on the grid to assure that only subtle changes in system frequency are necessary to continually match time and frequency.

INTERCONNECTED SYSTEM OPERATIONS

Now that we have covered the major building blocks of a power system (i.e., generation, transmission, substations, distribution, consumption, protection, and the elements of power grid organization), the next discussion explains the fundamental concepts, constraints, and operating conditions that make an interconnected power system stable, reliable, and secure.

Inertia of the Power Grid

Inertia is one of the main reasons large interconnected systems are built. Inertia is the tendency of an object at rest to remain at rest or of an object in motion to remain in motion. The larger the object, the more inertia it has. For example, a rotating body such as a heavy generator shaft will try to continue its rotation, even during a system disturbance. The more spinning generators connected together in the power grid interconnection, the more inertia the grid has available to resist change. Power systems boost stability and reliability by increasing inertia. Solar PV is an example of a generation type that does not contribute to grid inertia because it has no spinning objects.

The best way a power system can maintain electrical inertia is to have an interconnected system of several rotating machines. Note, the word "machine" is used opposed to generator because both motors and generators contribute to electrical inertia. Generation plants that do not have spinning members, such as solar voltaic plants, do not add to the system's inertia. The more inertia a power system has the better.

Power system stabilizers (PSS) are installed on generators to compensate for decreasing inertia under fault conditions. The electromechanical *governors* that control the amount of steam to the turbines, for example, are controlled by PSS during fault conditions to automatically oppose normal governor responses in order to maintain inertia and therefore help stabilize the system during an event.

Figure 8-5 illustrates the concept of inertia and frequency stability in a steady-state interconnected power system.

Suppose these trucks are carrying load and all are traveling at 60 mph. They are all helping each other go up the hill carrying the load. As the hill incline increases (i.e., system losses plus load increasing), the trucks must increase their throttles (i.e., governors) in order to maintain speed at 60 mph. If the incline got too great for these trucks to travel at 60 mph, additional trucks would have to be added and load

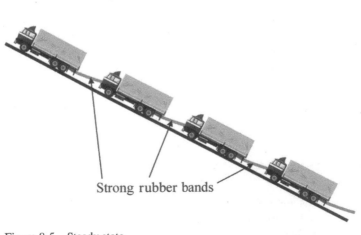

Strong rubber bands

Figure 8-5   Steady state.

distributed in order to maintain speed (i.e., frequency). As the incline decreases (i.e., less losses and load), the trucks must decrease their throttles in order to maintain speed. If significant load is removed, some trucks would not be needed and can be taken offline while the 60 mph speed is maintained. (The 60 mph is analogous to system frequency. The rubber bands are analogous to transmission lines. The trucks are analogous to generators. The trucks are carrying the load.)

In a large-scale integrated power grid, very similar concepts and actions apply. The grid generators are working together to share the load. Their electrical output frequency is a joint effort. They all slow down when load is added and they all tweak their rubber bands (i.e., transmission lines) when load changes. All generator units and transmission lines work together as a system to produce a highly reliable electric service that balances generation with load at a constant frequency and with good voltage.

Inertia comes into play when a fault occurs on the grid system. The sudden inrush of fault current causes the generators to slow down, high inertia helps the generators to keep spinning to push through the disturbance until the breakers clear the fault by their protective relays. Otherwise, voltage and frequency would collapse further.

Balanced Generation Conditions

Power out of the generator is a function of rotor angle. Zero power out has a zero rotor angle and maximum power out has a 90 degree rotor angle. When two same size generators are connected to one bus as shown in Figure 8-6, they are producing the same amount of power and their rotor angles are equal. This represents a balanced

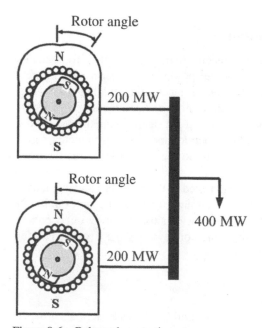

Figure 8-6   Balanced generation.

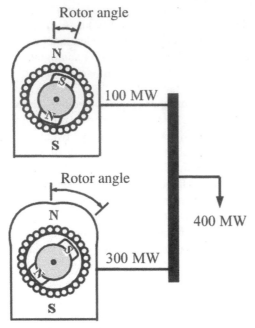

Figure 8-7   Unbalanced generation.

generation situation. Three generators would look the same but with more combined output power.

Unbalanced Generation Conditions

When two generators of the same size are connected to one bus and their rotor angles are not equal as shown in Figure 8-7, the output power of one generator is different from the other. This represents an unbalanced generation situation. Increasing the exciter current increases the rotor's magnetic field and produces a higher output voltage and more reactive power out. Increasing the steam to the turbine (prime mover) increases the real output power of the generator and also tries to increase grid frequency. Thus, increasing the exciter and steam to the turbine increases the rotor angle, power output, and system voltage.

Note, when two generating units are connected to the same bus and one unit is larger than the other, yet they output the same amount of power, the larger unit will have a smaller rotor angle than the smaller unit. Since maximum power out occurs at a rotor angle of 90 degrees, the larger unit would not have as great a rotor angle for the same amount of generator output power.

System Stability

Stability is the term used to describe how a power grid handles a system disturbance or power system fault. A stable system will recover without loss of load. An unstable

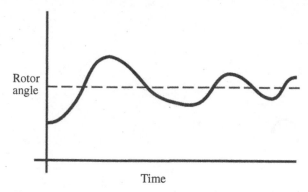

Figure 8-8   System stable.

system could trip generator units, shed load, and hopefully settle down with islands without a large-scale blackout.

System stability is directly related to generator loading. The generator's rotor angle changes when loading on the generator changes. As shown in Figure 8-8, a stable system that undergoes a system fault will have its generator rotor angle change/swing and then converge back to a stable steady-state condition. As long as the rotor angle converges back to stable, the system will eventually become stable. This is obviously a desired situation after a major line fault.

System Instability

Since the generator rotor angle changes when load conditions change, sudden large changes in generator loading can cause great power swings in rotor angle and create a condition of *system instability*. As shown in Figure 8-9, these great swings can cause the generator to become unstable and trip offline. Loss of generation causes underfrequency or overfrequency conditions on the rest of the system and unless generation-load balance is achieved quickly, load will be shed and outages will occur. Loss of load can cause more generators to trip as a result of excessive swings in their rotor

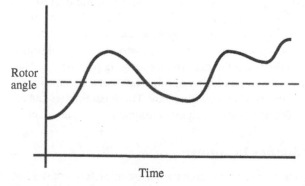

Figure 8-9   System instability.

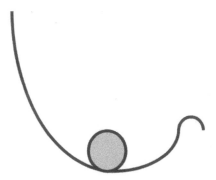

Figure 8-10   Conditional stability.

angles. The system will eventually become unstable unless something is done to re-establish balance between generation and load. Therefore, extreme load variations can cause a system to become unstable and possibly result in a wide spread outage or full system blackout.

Conditional Stability

Each generator unit and the grid as a whole normally operate in a condition called *conditionally stable*. For example, in Figure 8-10, if the ball is pushed up the wall to the left, it will roll back down to the bottom then to the right and hopefully settle back to the bottom. But if the ball is pushed up the wall too far and let go, it will actually keep on rolling up the right side and perhaps go off the edge, resulting in a generator tripping offline.

This analogy describes what happens to power system generators with regard to their rotor angles during system disturbances. Depending on the system fault (over-current condition), load breaker trips (undercurrent condition) or some other power disturbance that causes rotor angle instability, there is a conditional limit as to whether the unit or system will regain stability. Otherwise, generation and/or load breakers trip and the system becomes unstable resulting in cascading outages and possibly major wide area blackouts.

Most of the effort in analyzing power systems is deciding on the operational constraints and trying to determine the limits of conditional stability. The engineering and planning departments are constantly analyzing load additions, possible single, double, triple contingency outages, impacts of new construction, and all other planned or unplanned changes to the system to determine operating parameters. This engineering and planning effort tries to determine the fine line between maximum uses of system capacity versus stability after a contingency. These parameters can all change during peak and non-peak conditions, equipment maintenance outages, etc.

Unit Regulation and Frequency Response

A stable system is one where the frequency remains almost completely constant at the design value of 60 Hz. This is accomplished through unit regulation with quick frequency response. Only very small deviations from this standard frequency should occur. Generating units collectively control system frequency. Generators that are online as "*load following units*" usually provide the necessary unit regulation and frequency response actions that ensure the system is operating at 60 Hz at all times.

Note that electric utilities are always in a "*load following*" mode of operation. That is, consumers turn loads on and off at will, without notifying the utility. As a result, the utilities must adjust generation to the random changing load demands and predict/plan for future expectations.

SYSTEM DEMAND AND GENERATOR LOADING

Total system demand is the net load on the system within a controlled area that must be served with available internal generation and tie line import resources. Generators are put on the system according to their *incremental cost*, by the type of generator used, contribution to system stability, and other factors. Some generator types are designed as *base load units* that are capable of running 24/7 while others are designed as *load peaking units*. The load peaking units generally cost more to operate than the base load units. Another category of generator types is *load following units*. Load following units can be used as expensive base load units, can operate 24/7 but they are still typically not as expensive to operate as peaking units. Other generator types, such as wind are used whenever available.

A typical 24-hour demand curve showing internal generation requirements is shown in Figure 8-11.

This demand is supplied by base-load, load-following, and peaking generation units. Generator units 1, 2, and 3 are considered base load units (least expensive to operate and designed to operate 24/7). Generator units 4 and 5 are considered load

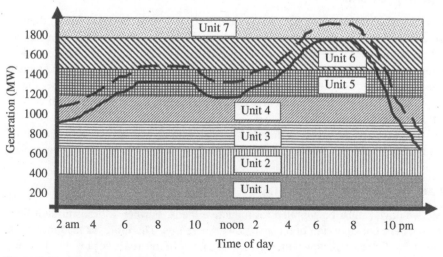

Figure 8-11   Generator loading.

following units (used to maintain ACE and tie line bias). Generator units 6 and 7 are considered peaking units (usually the most expensive to operate, yet can start quickly and help balance load with generation.)

Spinning Reserves

Normally, it takes several hours to restart a major fossil fired generator and sometimes days to restart nuclear plants after a trip. *Spinning reserve* is the term used to describe generation capacity that is readily available to go online almost immediately without operator intervention, should an online generator or an import transmission line trip due to a system disturbance. There are two types of spinning reserves; those necessary to meet changing load conditions and those that must respond quickly in the event of a disturbance. Generation units that meet changing load conditions are usually the "load-following" units. The other types of spinning reserve units are those that can respond quickly to help bring back system frequency and stability after the loss of a generator or import tie line. These quick response units can be originally offline peaking units providing they have "*fast start*" capability. However, spinning reserve requirements are set by criteria and standards published by the NERC. Generally speaking, combustion turbines are fast starting units that can be used for spinning reserves when certain criteria are met.

Normally *operating spinning reserves* are supplied by generation units that are already online meeting the changing load patterns. *Supplemental reserves* are units that are spinning but not serving load. Typically, units that account for between 5% and 10% of the load being served are also serving as spinning reserves. Other spinning reserve resources are peaking generators, combustion turbine generators, interruptible load, and lastly load shedding protection schemes are all used to restore system frequency following a disturbance.

Capacity for Sale

Since the generation owner has the option to *export energy* to other areas, there is also an opportunity to *sell excess generation capacity (energy)* on the spot market or through long-term sales agreements. The ability to make these sales is dependent on loading and available generation. For example, the northwest area of the United States usually has an abundance of hydroelectric generation for sale. Capacity that is above the utility's load requirement is excess generation capacity that could be sold on the open market.

Referring to the previous diagram (Figure 8-11), generator units 4, 5, 6, and 7 could run near full-load to provide energy for sale to other interconnected companies.

Reactive Reserves and Voltage Control

Reactive power must be supplied for inductive loads. Further, as transmission lines load up, their consumption of reactive power increases. Therefore, as load increases, the demand for reactive power increases. The supply of this reactive power must come from generation units, switchable capacitor banks, line charging, and tie line contract agreements. These reactive power resources must be readily available to the system operator in order to provide good voltage throughout the system. Power contracts outline the requirements for generation to supply both real and reactive power plus maintain proper voltage conditions. These resources can be shared at a cost. Therefore, real power and reactive power can be bought and sold on the open market, but must recognize the operating constraints to insure reliable system operations.

System voltage is controlled through the use of reactive supply resources, such as generation, switchable capacitors, and static VAR compensators. Voltage is controlled by switching on shunt capacitors and increasing generator output when system voltage is low and switching off shunt capacitors and switching on shunt reactors when system voltage is high. Usually the lowest system voltages occur during summer peak conditions when air conditioning load is maximum. Usually the highest system voltages occur during the very early morning hours when load is the lightest. Some areas have maximum load conditions during the winter months where resistive heating is maximum. Either way, the highest system voltage conditions usually occur late at night or early morning when load is minimum and the lowest system voltage conditions occur in the early evening when load is maximum.

Generator Dispatch

Generator dispatch is a primary function of day-to-day operations. The units on the system also include *customer owned, independent power producers* (merchant plants), aside from the standard utility-owned generation plants. Each has a cost or a contract requirement that must be considered in the dispatch arrangement. The operator plans the day by ensuring the lowest cost units are dispatched to "base load" criteria. Then higher cost units are dispatched as the load increases during the period. Other units may be required to "load follow" or for "peaking." The mix of base load and peaking units provide the system operator the resources necessary to effectively and reliably meet system demand with disturbance stability provisions in place.

The *required total generation* is determined by load forecasts plus any other applicable information available that would or could affect operations (such as weather, emergencies, equipment out of service for maintenance, and other seeable and unforeseeable factors).

Load Forecasting

Accurate and complete load forecasting has become a valuable factor in day-to-day power grid operations, future projections for generation and facility requirements, outage, and contingency planning to name a few. Load forecasts are used to schedule power flow on transmission lines and when it is time to change out underrated equipment. Load forecasts are used in demand side management systems for curtaining load during energy emergencies. They are also used for budget planning, revenue forecasting, and operational needs assessments.

Utilities share load forecasts for determining timing on joint venture projects, sharing generation resources, providing adequate facilities for preventing major system disturbances when faults and outages occur, and the list goes on and on.

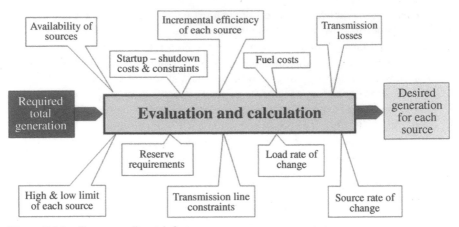

Figure 8-12 Generator dispatch factors.

Short-term or hourly load forecasts help balancing authorities provide dependable generation–load balancing procedures. Load forecasts must be adjusted for weather expectations in order to provide a stable operation projection and to correctly track the ACE equation on a real-time operating basis.

The process of deciding which units to use to meet a daily or weekly requirement is extremely complex. There are many variables that must be considered. To help solve this problem, many utilities use a program called *unit commitment*. The diagram in Figure 8-12 illustrates some of the factors that go into deciding what units should be used or committed to meet a load forecast.

RELIABLE GRID OPERATIONS

First, let us review the terminology differences regarding planned and unplanned outages, equipment failures, faults, and other events that, if not properly controlled, could lead to cascading outages and possible system wide blackouts. For example, high-voltage equipment, such as transformers, circuit breakers, and regulators require periodic maintenance. Scheduled or planned outages are required to isolate such equipment in a manner that allows for safe maintenance activities; this is usually accomplished by careful switching operations. However, such equipment can fail unexpectedly causing high short circuit or fault current to flow back to sources causing voltage and frequency issues until the protective relays sense the undesirable condition and send specific control signals to trip open associated circuit breakers. Faults are cleared by protection devices (i.e., breakers, fuses, etc.) and can also cause customer outages in the meantime. Outages are referred to as the loss of electrical service to customers by either planned maintenance operations or circuit breakers trips due to power faults on equipment or lines. Outages are normally associated with areas without power or number of customers without power.

During normal operations, substation equipment, transmission lines, or distribution feeders that need to be taken out of service for any reason not associated with protective relay trips would be considered planned or scheduled outages. In this case, electrical service is typically re-routed so that consumers are unaffected by the scheduled maintenance activities. There can be several lines or substation equipment out of service at any given time, as long as the interconnected system and customer service remains intact and reliable.

In comparison, emergency operations occur when there are unexpected equipment failures or line outages due to protective relay or fuses operations and either consumers are out of service or redundant equipment is in place that is able to maintain electrical service until repairs can be made. In the more extreme events, equipment outages lead to overloaded condition of other equipment that cause additional outages (the "domino effect"). Sometimes the outages affect the generation-load balance, where cascading outages can occur that could lead to a large-scale outage and possible system blackout. System blackouts require emergency restoration procedures usually starting with a power plant that has blackstart capability. Blackstart capable power plants activate emergency generators that start the plant process working again; then transmission lines are energized to critical substations, while picking up load in small increments, and creating transmission *"cranking"* paths to enable other power plants to startup and come back online. Eventually, the interconnection is restored and all customers are again provided electrical service.

Factors that contribute to the reliable grid operations are discussed below for both normal and emergency operating conditions.

Normal Operations

Normal operations occur when all load is being served with stable frequency, good voltage, proper transmission line flows, ample reserve margins, and little known activity that could suddenly grab the attention of the system operator to take immediate action. In today's environment "normal operations" mean operating several generation units and transmission lines at or near full capacity, trying to schedule equipment out of service for maintenance, and responding to daily events, such as planned outages, switching lines and equipment for maintenance, and coordinating new construction projects.

The behind-the-scenes day to day normal operations will now be discussed.

Frequency Deviation

Generators are limited to a very narrow operating bandwidth around the 60 Hz frequency constraint. *Frequency deviation* within an electric system outside tight parameters will cause generators to trip. Since transmission systems are interconnected to various generation sources, excessive frequency deviation may also trip transmission lines in order to protect sources of generation supply.

Frequency deviation must be carefully monitored and corrected immediately. The system operator is watching for the common causes of frequency deviation conditions, such as:

Sudden Supply/Demand Imbalance   Loss of supply can reduce frequency. Loss of load can increase frequency. Either way, frequency deviation is not tolerable and the system operators and generator operators must be ready to take immediate action upon loss of generation.

When a unit trips and the load-generation balance is upset, the restoration process involves automatic and then manual intervention to *"arrest frequency."* When a generator trips in a specific area and system frequency drops, spinning inertia immediately helps stabilize frequency, then unit governors and PSS bring rotor angle swing under control (usually to a frequency below 60 Hz), then AGC signals start to bring up generation toward the normal 60 Hz frequency. Finally, the balancing area having the generation outage brings on spinning reserves to correct the ACE to where the frequency is restored back to 60 Hz. These steps ultimately restore frequency unless the disturbance is great enough to cause a wide area outage that results in cascading outages (domino effect) and the grid separates into islands of independent flat-frequency control.

Short Circuits or Line Faults   Faults on major transmission lines are usually cleared by opening circuit breakers. Line outages change power flows throughout the grid and can suddenly add load to generators and/or suddenly remove load from generators. Some lines and transformers can become overloaded, requiring system operators to take further action or chance other breakers tripping to clear overloaded lines or equipment.

Initial emergency response during these conditions is often automated, however manual intervention comes into play as soon as the operators can assess the situation and implement additional corrective action. The system control operator is standing by to take remedial action, should an abnormal event occur suddenly. Operator situational awareness is key to circumvent any possible line overload, voltage collapse, or frequency deviation condition.

Cascading Failures

Cascading failure situations may be created by any abnormal condition or system disturbance. They can result in the loss of transmission and/or generation in a cascading sequence. For example, the August 2003 outage that affected most of the Northeastern United States was due to cascading outages. The scenario began by having some transmission and generating facilities in the Northeast out of service for maintenance. Then, one of the remaining transmission lines in service tripped because it sagged into a tree under heavy load conditions. At the time of the trip, major cities in Ohio were in a "heavy" import condition meaning much of the energy was being supplied by the transmission interconnection system. Once the first line tripped, the remaining interconnected transmission system started to overload and one by one several major transmission lines tripped off line.

As the transmission lines began to disconnect, the system experienced sections having excess load and shortage of supply. This created a frequency deviation situation and the remaining online generation began to slow down due to the overload condition. The utilities involved did not have adequate generation reserves online at the time to meet the sudden demand, therefore generator units began tripping. As generation tripped, the problem continued to worsen.

Each system at the time of the initial failure had some time and opportunity to *island* (i.e., separate from the grid) once the supply was inadequate to meet the load. This time frame for the control operators was probably less than a minute. In that time frame, if the utility did not "disconnect" from the grid and was not able to meet internal load with reserves or through an operational underfrequency load shedding scheme, the utility remained on the grid and the cascading failures continued.

Eventually, the entire grid was left with no supply. Only those systems that disconnected were able to survive, at least partially, this cascading failure scenario. Unfortunately, the U.S. grid has seen an increasing number of these failures over the last several years, due to delays in building more power plants and transmission lines. Cascading failures can be prevented. The following changes can improve system reliability to reduce the possibilities of future cascading disturbances:

Construction of New Resources As the need for more electrical energy increases and as a result of restructuring, adequate resources in the form of additional transmission lines and generation have not kept pace. This has resulted in lower reserve margins for many utilities. Building more generation plants and transmission lines will significantly improve system reliability. The country is, however, implementing several new special protection systems and other smart control schemes to help improve grid reliability.

Transmission Ratings NERC has recently undertaken efforts to re-rate transmission facilities and dictate when facilities can be taken out of service. Making sure there is adequate transmission capacity at all times or rating the lines so that import limits are set to maintain system integrity is essential to system stability.

Underfrequency Shed Schemes Utilities are now required to update these protection schemes to ensure they are adequate to meet the new load and grid requirements.

Control Operator Training New guidelines and requirements are in place to ensure operators are certified and have continuing education and training to keep pace with changes in system requirements.

Voltage Deviation

Voltage on a system can deviate and cause system operational problems. Voltage constraints are not as restrictive as frequency constraints. Voltage can be regulated or controlled by generation or other connected equipment, such as regulators, capacitors, and reactors. Usually the equipment served (i.e., load) is less sensitive to voltage fluctuations than frequency. The control operator is watching for any of the following conditions to occur that would cause system voltage to deviate substantially:

Uncontrolled Brownout An uncontrolled *brownout* is a condition where excessively low voltage is experienced on an electric grid. This condition can persist for long periods of time and can result in equipment failure (i.e., motors or other constant power devices). Some loads, such as lighting and resistive heating, might show flicker or heat reduction from low-voltage conditions but not become damaged.

Voltage Surge   *Voltage surges* usually result when services are restored and high/low voltage transients occur. Voltage surges are usually transient or short term in nature. This type of voltage deviation may damage consumer equipment (like computers) and possibly lead to other equipment failures.

Normally, utilities are required to maintain voltages within tolerances set by industry standards or regulatory authorities. Manufacturers are expected to design consumer equipment such that it can safely operate within normal power company service tolerances. System operators are responsible for preventing deviations that exceed specified tolerances. System operators are also expected to ensure voltage stability through constant monitoring and adjustment of the system's real-time conditions.

Emergency Operations

Emergency operations exist when the power system is experiencing outages, faults, load shed, adverse weather conditions, voltage, and/or frequency instability. These problems or conditions require the immediate attention of all operating personnel.

Planning and general operating criteria established by regulatory agencies and individual utility companies try to insure that the system remains stable under a variety of normal and abnormal conditions so that emergency operations can be avoided. Operation of any electric system during abnormal or emergency conditions requires specially trained and highly experienced operators. Often the operator's experience and familiarity with the system capabilities can mean the difference between a small area disturbance and a total system shutdown. This section deals with various conditions and typical operating guidelines imposed on operators under emergency operations conditions.

The behind-the-scenes emergency operations will now be discussed.

Loss of Generation

Equipment failure, load imbalance, or other malfunction can cause a generator to trip offline. This loss of generation will result in more load than supply until the situation can be resolved. Since electricity is not stored, the power system reacts to the difference between generation and load by a change in frequency. The response is immediate and requires corrective reactions within a very short time frame. To compensate for loss of generation, the following planning criteria are in place:

Spinning Reserve   *Spinning reserve* as discussed earlier provides additional generation online and ready to accept load. The typical requirement for spinning reserve is "5–10% of load, being served or loss of single largest contingency." If, for example, the generator that trips is the largest unit on line, the utility should have access to spinning reserves that will compensate for the loss of the unit. However, putting spinning reserves into play requires some reaction time.

Transmission Reserves   *Transmission reserves* can provide instantaneous response to loss of generation. Operators carefully monitor transmission loading conditions and available capacity in the event transmission reserves are needed.

Emergency Generation   There are some systems where emergency generation can be started in a short time frame (10 minutes or less). The interim period may be handled by a combination of spinning and transmission reserves. Emergency generation, such as peaking units are usually located in substations and are fueled by diesel or another fuel source that can be easily replenished.

Controlled Brownouts   If the mismatch between generation and load is not too great, it may be possible to compensate by reducing distribution voltages. This condition is called a *controlled brownout*. When this condition occurs, lighting dims slightly (sometimes not noticeable). The reduced voltage results in less power being consumed by resistive loads (such as electric heaters, incandescent lights, and other resistive residential or business loads).

Rolling Outages   If there is a lack of spinning reserve and transmission capability, and if the utility cannot bring supply up to meet load quickly, load shedding is the only option available to ensure the system remains stable. This approach is typically referred to as a *rolling blackout*, or just *blackout*, where operators trip and close substation distribution breakers. Underfrequency protective relays automatically trip distribution breakers during underfrequency degradation conditions. However, operator intervention of load shed breaker tripping allows frequency to remain stable before the underfrequency relays start automatically tripping load. This approach is usually a last resort as it does result in loss of revenue and lower customer satisfaction ratings.

The impact of loss of generation and the resulting emergency operations are dependent on the utility's generation and transmission resources. A utility that is highly dependent on generation is susceptible to constraining conditions for the loss of a unit. A utility that has the majority of its energy provided by purchases from other utilities over transmission tie lines usually experience less chance of losing their own generating units. However, they are more dependent on system disturbances and uncontrollable events that are outside their system.

The reliability criterion established by NERC requires that the utility or controlling party adjust system parameters within 10 minutes after a loss of generation in order to prepare for the next worst-case contingency. Ten minutes is not much time because another event (such as another relay trip operation) could occur in the meantime.

Loss of Transmission Sources

Losing a major transmission line due to weather or malfunction is much the same as loss of generation. Since the transmission system delivers energy in both an import and export mode, loss of a transmission line may result in different scenarios.

Export   The loss of a major transmission line when the control area is in export mode results in too much generation for the local load being served. Without correction, the system could experience severe over-voltage and/or over-frequency conditions. The over generation that is causing high voltage and frequency can be rectified by reducing local generation.

Import The loss of a transmission line for a control area in the import mode results in an excess of load compared to supply. This scenario is identical to the loss of generation where the system frequency and voltage decrease. Automatic load shedding schemes try to balance load with available generation. Outages are still possible. As internal generation comes online, load is restored.

Due to increased restrictions on generation, many utilities are dependent on transmission sources to meet growing energy demands. Often times, the loss of a transmission line is more serious than the loss of a generator.

CHAPTER 9

SYSTEM CONTROL CENTERS AND TELECOMMUNICATIONS

CHAPTER OBJECTIVES

After completing this chapter, the reader will be able to:

- ☑ *Explain the functions and equipment of Electric System Control Centers*
- ☑ *Describe how SCADA (Supervisory Control and Data Acquisition) enables remote control of substation equipment*
- ☑ *Explain the functions of the Energy Management Systems*
- ☑ *Describe the EMS software tools used by system operators*
- ☑ *Explain how synchrophasors and Wide Area Monitoring Systems help system reliability and security*
- ☑ *Describe the types of telecommunications systems used by power companies*
- ☑ *Discuss how advances in digital substation equipment and automation impacts efficiencies in system control, equipment wiring, and functionality capabilities*

ELECTRIC SYSTEM CONTROL CENTERS

Electric System Control Centers (ESCC) like the one shown in Figure 9-1 operate 24 hours a day, 7 days per week making sure the electric power system within their control area is operating properly. System operators monitor their control area looking for signs of possible problems and taking immediate action to avoid major system disturbances, should a warning sign occur. Operators are tasked with the responsibility to maintain system connectivity, reliability, stability, and continuous service. They are also responsible for coordinating field crew work activities (i.e., clearances) to make sure crews are safely reported on high-voltage lines and equipment. System control center operators have noteworthy responsibilities.

Under normal conditions, control operators monitor the system and are prepared to respond immediately to incoming alarms from equipment out in the field. Under emergency conditions, control operators respond cautiously to incoming

Electric Power System Basics for the Nonelectrical Professional, Second Edition. Steven W. Blume.
© 2017 by The Institute of Electrical and Electronics Engineers, Inc. Published 2017 by John Wiley & Sons, Inc.

Figure 9-1   Electric system control center.

alarms, requests from field personnel, and inter-agency communications alerts. They realize the complexity of controlling a major system and the possible consequences, should they make an error in judgment. They are highly trained on situational awareness and ready to take immediate action.

System control operators have many tools at their disposal. These tools help them look ahead if something were to happen, analyze "what if" scenarios based on real-time loads and line flows, and they have direct communication lines to people in other strategic locations.

The main tool of the ESCC operator is the *Supervisory Control and Data Acquisition* (SCADA) system. This system allows control operators to monitor equipment conditions, control equipment as necessary, dispatch generation, and obtain written reports of all parameters about the power system. The SCADA system is made up of a centrally located *master computer* and several *remote terminal units* (RTUs) located throughout the system. An equipment failure or breakdown in the telecommunications equipment supporting SCADA can cause control operators to make incorrect system adjustments. For example, a communication channel between the master computer and an RTU would not update the operators' information about the status of a substation equipment operation. The operator would not know if a breaker were actually open or closed. The lack of up-to-date information is detrimental to the reliable operation of the system, especially during disturbances when critical decisions have to be made.

Telecommunications equipment is used to communicate information electronically between the ESCC and the several RTUs. When problems occur in telecommunications equipment or control center equipment, system operators must occupy *back-up control centers* in order to resume monitoring and control functions of the power system. Control centers and back up control centers normally have emergency generators and *uninterruptible power supply* (UPS) systems to make sure computers, lights, communications equipment, or other critical electric dependent loads are powered without interruption.

This chapter discusses the equipment used in ESCCs, RTUs, and telecommunications. Upon completion of this chapter, the reader should have a fundamental understanding of what is involved in system control operations.

SUPERVISORY CONTROL AND DATA ACQUISITION (SCADA)

The basic operation of virtually every electric utility in the United States now relies upon SCADA systems. Up until the late 1940s, many utilities had personnel stationed at substations. In some cases, these were residents who remained on call 24 hours a day. With the advent of SCADA systems, it was no longer necessary for utilities to maintain manned operation of substations. Additionally, utilities need access to system information immediately to properly control the power system.

The basic function of the SCADA system is to remotely control all essential equipment in each substation from a single control center or backup control center. The functions in the substation that are communicated to the control center are to measure, monitor, and provide control of all critical substation equipment. At the control center, the basic functions are to display and store the information, generate alarms if anything abnormal occurs, and to enable remote control operation of equipment in the substation to initiate changes in the effort to regain normal operation. Also, other equipment not found in substations might have remote control capability through SCADA, such as backup control centers, transmission line motor operated switches, emergency load transfer switches, demand side management (DSM) automation, and other communications equipped control and monitoring devices.

SCADA systems have the capability of providing graphical representation of the generation stations, transmission lines, substations, and distribution lines. Depending on control area responsibilities, ESCC operators have full control of their control areas and responsibilities. They might also have monitoring only capability of equipment and lines in adjacent control areas of the interconnected system.

SCADA alerts a control operator that a change of state occurred. Usually SCADA gives the operator full control of operating equipment to change the state back to normal. If an operator closes an open breaker via SCADA for example, then SCADA will in turn alarm the dispatcher that the breaker status is now closed. This feedback indication technique is inherent in the SCADA system. This allows operators to verify actions actually have taken place and the operator can monitor results afterward.

SCADA systems normally update information about every 2–4 seconds, the amount of time it takes the main computer to exchange information to each RTU sequentially. In reality, a lot can happen in between scans and data updates. During normal operations, this time delay is tolerable. However, during disturbances or emergency situations this time is long. The use of *synchrophasors* (also called phasor measuring units or PMUs) in conjunction with *Wide Area Monitoring Systems* (WAMS) is significantly reducing this time lag. The use of PMUs and WAMS are discussed in more detail later in this chapter.

Figure 9-2 outlines the equipment that comprises a SCADA system, including the control center, RTUs, and telecommunications equipment. Notice the map board, the main computer, and the various communications systems that connect RTUs to the main or central computer.

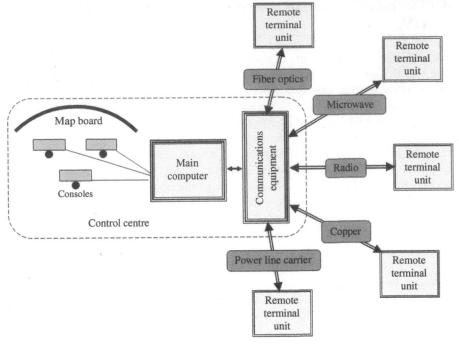

Figure 9-2   SCADA system.

Data Acquisition Functions

The *data acquisition* portion of SCADA gives operators the ability to remotely monitor analog electrical quantities, such as real/reactive power, voltage, and current in real time. Also, operators are alerted to problems as they occur through alarm and indication points. For example, tripped breakers, security breaches, fire alarms, and enunciator alarms send signals to the control center where a visual and/or audible alarm attracts the attention of the system operator. The operator then makes changes remotely with the control functions of SCADA or informs other operating departments that equipment might require inspection (such as low oil level alarm on a transformer).

Some examples of analog data acquisition information include:

- Bus volts
- Transformer watts
- Feeder amps
- System VARs
- Regulator position
- Entry/security alarms

Examples of alarm and indication information include:

- Breaker 1274 now open
- Motor operator switch 577 now closed
- Station service power now off
- Control building door now open

Also, SCADA enables the communication of accumulation data such as the following:

- Generator unit 1 MW-hours
- Generator unit 1 MVAR-hours

Control Functions

The *control* portion of SCADA allows operators to remotely control/operate equipment at a particular substation such as:

- Close breaker 1274
- Open motor operator switch 577
- Start emergency generator
- Circuit breakers

ENERGY MANAGEMENT SYSTEMS

Energy management systems (EMS) became a major extension from SCADA with the arrival of advanced computer programs and applications. Sophisticated computer programs were developed to monitor system conditions in real time and initiate automatic programmed control responses to assist the operation of actual equipment. The perfect example of an automatic power grid function that is controlled by many EMS systems is generation. *Automatic generation control* (AGC) is the most comprehensive development of EMS in use today. Smart computer programs are used to ramp up and down generators based on best economics and system reliability factors.

Other very important EMS management computer program tools were developed to improve the reliable operation of large interconnected power grids. These software tools help reduce power production costs, improve real-time analysis of current system operating conditions, provide information to avoid wrong decision making by operators, improve system reliability, security, and much more. The umbrella term used to describe all these important system operation software tools is known as the *energy management system* or EMS.

The most significant EMS software programs in use today are described below:

State Estimation

The *state estimator* uses a computer model of the system, taking into account line outages for maintenance, etc. and uses almost real-time SCADA information to display to the operators all line and substation equipment power flow, voltage levels, and equipment status. The information is presented in a way that helps operators see the status situation of the system at a glance; usually in terms of capacity utilization percentages. The state estimator calculates bus voltages and power flows where sensor information (CTs and PTs) is not available. The state estimator uses all available sensor measurements, known facts about the system (equipment outages), and other relevant information (weather) to calculate the best possible estimate of the true status (or "state") of the power system. For example, the state estimator is used to calculate new power flow conditions, such as voltages and currents when lines are taken out of service in order to help system operators predict "what if" scenarios.

Contingency Analysis

The reliability software programs in the EMS perform "what if" scenarios to determine worst-case problems that might result if each major line or transformer were taken out of service for any reason. The output ranks potential contingencies according to severity and probability of occurrence, and lists what recommended actions system control operators can take, should such events occur.

Further, if a line is to be taken out of service for any reason, the contingency analysis program determines the next worst-case contingency scenario so that operators can adjust the system to best handle the next contingency.

Transmission Stability Analysis

The reliability software runs a series of outage scenarios based on real-time conditions looking for transmission line loading conditions and other system short comings that can push the system close to stability limits. It looks for increasing voltage violations, increasing VAR requirements, interchange transactions that can cause problems and reports those results to the system operator or ESCC engineer.

The software also looks for voltage stability issues to avoid low voltage and voltage collapse problems.

Dynamic Security Assessment

To help system operators identify other potential problems, the *dynamic security assessment program* reports system equipment that is reaching rating threshold conditions in real time. For example, bus voltage approaching over-limit, lines approaching over loading, etc. are reported to the operator. It also takes into consideration thermal constraints and emergency ratings. This helps operators identify potential problems before they happen and helps by providing operating margins during emergency conditions.

Emergency Load Shedding

The EMS is capable of shedding load in an emergency. Similar to underfrequency load shed relays, the EMS can trip load fed from circuit breakers if the frequency declines. The operator can drop load fast and effectively via this system. System operators can coordinate rolling blackouts before the automatic load shedding relays operate.

Power Flow Analysis

Static information about the system's lines, transformers, etc. is entered into the computer programs regularly. For example, conductor resistance of a new transmission line that is scheduled to go into service is entered into the EMS data base. The EMS then calculates the new *power flow* conditions with the new line included. The software can report detail system information during daily, weekly, monthly, and yearly peak conditions. This power flow data are very useful to planning engineers to determine future power system additions.

Generation Planning, Scheduling, and Control

The EMS is an effective place to plan generation needs. This planning software incorporates load forecast information, generation schedules, interchange, or tie-line exchange schedules, unit maintenance schedules, and unit outage situations to determine the best overall generation implementation plan. Further, based on all these schedules, the AGC part of the EMS actually controls the dispatch of generation. System operations, area control error (ACE) and frequency are then monitored according to this schedule to assure system reliability and compliance.

Economic Dispatch

The *economic dispatch* software allocates available generation resources to achieve optimal area economy. It takes into consideration generator incremental loading costs on an individual generator basis, transmission line losses, and factors in reliability constraints. The EMS helps determine *optimal power flow* based on actual generator data, contingency constraints, and real-time loading.

Reactive Power Scheduling

The EMS has the capability to schedule (usually up to 24 hours ahead) the controllable reactive resources for optimum power flow based on economics, reliability, and security.

Dynamic Reserves Analysis

The EMS can periodically calculate the reserve requirements of the system. For example, spinning, 10 minute, 30 minute predictions are made for a close look at generation requirements and resources. The program takes into consideration operating circumstances (i.e., largest unit online and timeframe requirements to make changes) to generate reports and alerts operators and engineers if necessary.

Load Profiling and Forecasts

The EMS software has the ability to produce load forecast reports. For example, next 2–4 hours on a running basis or next 5 or 7 day forecasts on an hourly basis can be performed by the EMS. These forecasts take into consideration weather information, history trends, time of day, and all other variables that could affect system loading.

Demand Side Management

As discussed earlier in Chapter 6, DSM is used to reduce load during certain on-peak conditions. The control signal used to shed interruptible load comes from the EMS. The EMS's DSM program decides when to initiate the broadcast signal that results in effective load reduction. The conditions for which signal broadcasting are required is programmed into the decision logic of the EMS.

Energy Accounting

Since all the records of sales, purchases, meter readings, and billings are centralized in the EMS data base; energy accounting reports are generated for management and the regulatory authorities.

Operator Training Simulator

The EMS has the capability to have a functioning operator training console that can be put into real operations at any time. The *training simulator* gives power system operators real experience using real system terminology, labeling, and electrical quantities on a real-time basis. However, the actual control points are deactivated to the trainee.

WIDE AREA MONITORING SYSTEMS (WAMS)

As we discussed earlier in this chapter, system reliability and security is primarily dependent on providing good voltage and stable frequency while demand changes. Transmission system operators and computer programs focus on making sure good voltage is provided throughout the interconnection (or power grid) to avoid voltage collapse issues that can lead into major disturbances. The generation system operators are focused on making sure the 60 Hz frequency is stable by assuring that a stable balance is maintained between load and generation. Automatic tripping of load breakers or generation units can occur if this generation-load balance goes outside of tight parameters.

Now there is a third parameter that system control operators and computer programs can monitor that further helps assure system reliability and security. The use of *Synchrophasors*, also called "*Phasor Measurement Units*" (or PMUs), in conjunction with a WAMS can now be used to identify potential grid disturbance issues in real time much faster than traditional SCADA and EMS systems.

WAMS is a relatively new approach to help maintain and assure system reliability using remote PMU sensors, telecommunications, and *phasor data concentrators* (PDCs) in centralized monitoring centers. The PMU sensors are installed in various substations throughout the grid interconnection, where the voltage (or current) phase angle information is time tagged using the GPS (Global Positioning System) and collected at a centralized location. The combined data information is analyzed for possible pre-emptive action, should system adjustments be necessary to avoid a major system disturbance.

In essence, WAMS, with strategically located PMUs present system operators with real-time information about the twist across the grid. Choosing a reference point on one phase, at one location in the grid to represent when the voltage sine wave crosses the zero axis, for example; say Atlanta, Georgia in the Eastern Interconnection, and comparing the phase angles' lag or lead of other zero crossings throughout the grid interconnection states the voltage angle twist across the grid. Should a suspiciously large voltage phase angle difference occur in the grid, actions can be taken to bring this twist back to acceptable levels far before a system disturbance occurs. This data collection and analysis happens in essentially real time. Whereas; SCADA and EMS have an inherent time delay of approximately 2–4 seconds; often too late to take corrective action, should a disturbance be imminent.

In its present form, a WAMS may be used as a stand-alone infrastructure that complements the SCADA system. In the future, WAMS technologies are expected to be incorporated into actual grid control systems.

Synchrophasor Technology

Synchrophasors or *PMUs* compare voltage and/or current waveforms across the grid. Figure 9-3 shows how PMU information can track the voltage twist in an interconnection. Notice the reference bus "E" where the voltage phase angle is said to be "zero". In other words, reference is made to the peak of the voltage waveform at bus "E". Now notice how all the other buses have either lagging or leading voltage phase angles.

Should the voltage phase angle increase across the interconnection to concerning magnitudes, the system operators and generation operators can change generator excitation and rotor shaft speed to bring down the voltage angle difference to acceptable levels.

Most systems today have synchrophasors in service and data are being collected that will eventually be used to provide this third dimension to system reliability and security control. Phasor data and application software are valuable assets for grid reliability because SCADA does not provide grid operators and planning engineers real-time information and wide-area visibility as to what is happening across a region or interconnection.

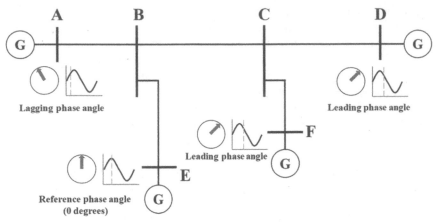

Figure 9-3   Synchrophasors in WAMS.

SMART GRID

The modernization of the electric power grid, from generation through consumption, is undergoing substantial change. All aspects of the power grid are becoming more digital, automated, efficient, informative, responsive, adaptable, secure, reliable, and other advancements as time goes on. The original power grid is transitioning into what is being termed "*smart grid*". Smart grid is the integration of all available power facility infrastructure with enhanced information and control capability, that incorporates systems automation, advanced telecommunications technology, and comprehensive solutions to improve power reliability and consumption optimization. The improved utilization of assets, or asset optimization helps delay additional infrastructure investments while giving consumers more control and utilization options.

Smart grid encompasses all aspects of the power grid and its load, including substation automation (primarily microprocessor adaptive relays, intelligent devices for auto transfer schemes, digital sensors, merging units, LAN Ethernet communications, etc., discussed earlier), transmission automation (primarily PMUs along with wide area networks were discussed earlier), and distribution automation (discussed next). Note: DSM, home energy management, and advanced metering infrastructure are considered part of smart distribution. Therefore, the remaining topic to discuss under the umbrella of smart grid is "smart distribution."

Smart Distribution

Smart distribution involves the redesign of the power delivery and consumption infrastructure to incorporate modern advances in distribution automation, customer information systems, digital electronic equipment, and distribution management.

Smart distribution has encouraged utilities to add more remote controlled devices, such as motor operated air disconnect switches, sectionalizers (i.e., reclosers without reclosing activated), capacitor voltage control devices, *programmable logic controllers* (PLC) for enhanced service restoration and several other technologically innovated power system devices and computer software programs that enable fast fault detection, failed equipment isolation, automatic sectionalizing, and service restoration that improves service reliability. Figure 9-4 provides an overview of smart distribution equipment and connectivity footprint.

Smart distribution is rapidly improving customer service reliability by reducing the average *number of outages* customers experience and their average *outage duration*. Advanced customer information systems that are connected to this distribution modernization offers a renewed level of quality of service. A summary of these new smart distribution and automation systems is provided below:

Distribution Automation

- Smart devices connectivity (protection equipment with digital interfaces)
- Digital communications efficiencies (incorporating Ethernet in substations)
- System operational performance and reliability enhancements (faster outage restoration)
- Enterprise enhancements (data storage, reporting, alerting, etc.)

Feeder Automation

- Improve SAIDI, SAIFI, etc. (industry standard reliability indices)
- Power quality improvement through better monitoring with automatic action devices, such as switched capacitor banks, and regulator controls

Distribution Management Systems

- Customer Information Systems (CIS)
- Geographical Information Systems (GIS)
- Outage Management Systems (OMS)
- Asset Management Systems (AMS)
- Advanced Metering Infrastructure (AMI)

Advances in Real-Time Operations

- Topology processing and mapping
- Integrated capacitor voltage control techniques
- Fault detection, automatic isolation, and service restoration
- Distribution load flow analysis for better planning
- Distribution state estimation and disturbance prediction

Advances in Offline Operations Planning

- Optimal feeder reconfiguration studies
- Optimal capacitor placement
- Optimal voltage profiles

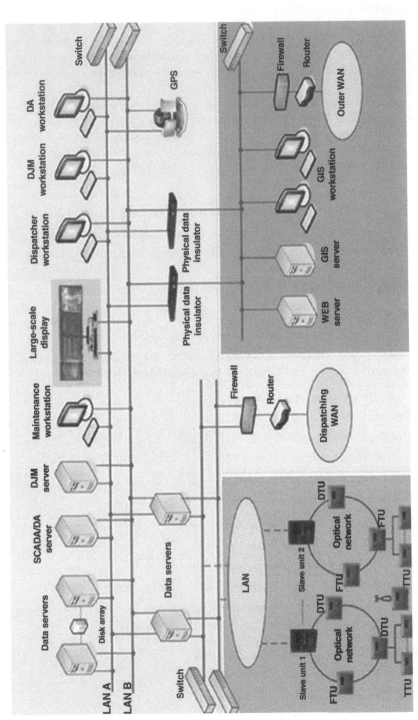

Figure 9-4　Smart distribution. Courtesy of NRelectric.com.

Outage Management

- Fault detection, isolation, and service restoration (FDIR)
- Integrated Volt-VAR control (IVVC)
- Optimal feeder reconfiguration (OFR)
- Distribution power flow (DPF)

Customer Automation

- Advanced metering infrastructure enabling two-way communications between the utility and consumer
- Enables home EMS
- Appliance control
- Small renewable energy generation integration
- Electric vehicle integration

Telecommunications

Telecommunications systems play a very important role in the reliable operation of large interconnected electric power systems. Advanced high-speed data networks are used for SCADA, system protection, remote metering, corporate data, and voice communications. Modern equipment like that shown in Figure 9-5 is used to provide communications services for customer call centers, service center dispatch operations, corporate voice lines, system control center private lines, direct interagency communications circuits, analog modem channels, and other services. Video networks are used for security surveillance, video conferencing, and enhanced training programs. These electronic communications networks are normally designed, built, and maintained by the electric utility.

Figure 9-5   Communications equipment.

These data, voice, and video networks are generally made up of six distinct communications system types, as follows:

➤ Fiber optics

➤ Microwave

➤ Power line carrier

➤ Radio

➤ Leased telephone circuits

➤ Satellite

The fundamentals of each of these communications systems are discussed next.

Fiber Optics

Fiber optic communication systems are being installed on electrical power systems all over the world. They are used for a host of services. The majority of the applications are for electric operations and a considerable amount of fiber is used for customer products and services. Additionally, fibers are leased to third parties as another source of revenue to the electric company.

Generally speaking, a fiber cable can have as few as 12 fiber stands or as many as 400 plus fiber strands depending on need and cable type. The photo in Figure 9-6 shows overhead optical static wire (*optical ground wire,* OPGW) coiled and

Figure 9-6   Substation fiber optics.

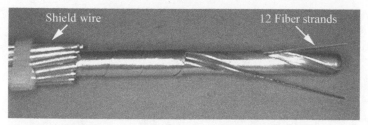

Figure 9-7 OPGW.

terminating in a substation. The OPGW is connected (spliced) to a non-conductive all *dielectric fiber cable* going into the control building.

The photo in Figure 9-7 shows a piece of OPGW. Lightning does not damage the optical fibers because fiber is made of non-conductive glass. Note the two buffer tubes of 12 fiber strands each that are contained in the center of this OPGW cable.

A fiber strand is made up of a very small glass core (approximately 8 micrometers in diameters), a glass cladding around the core (approximately 125 micrometers in diameter), and a color coding acrylic coating around the cladding (approximately 250 micrometers in diameter). The acrylic coating adds identification and protection.

Light pulses are transmitted into one end of the fiber strand core and exit the opposite end of the fiber strand core. The light pulses reflect off the surface interface between the core and the cladding based on the principle of *reflection of light*. Reflection of light is the principle that makes one see mountain reflections in calm lakes. The light pulses exit the fiber core slightly wider than when they entered the fiber. The longer the fiber is, the wider the output pulse becomes. There is a practical limit as to how often pulses can enter the fiber so that the resulting output pulses do not overlap, thus prohibiting accurate on/off detection. Typically, fiber optic distances reach 100 km without repeaters. Figure 9-8 shows how pulses enter and exit a fiber strand. The light must enter into the core within the *aperture angle* in order to enable the reflection of light to occur between the core and cladding. Sharp bends in the fiber will not let reflection of light occur and data errors result.

Electronic on/off digital communications signals are converted into on/off light pulses using fast responding laser diodes. The laser is pointed into the core of the fiber. At the receive end, a very sensitive fast responding photo detector transforms the optical pulses back into electronic pulses for the communications equipment to

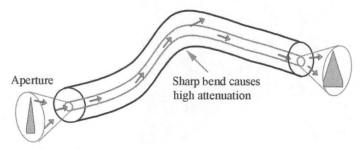

Figure 9-8 Fiber optic principles.

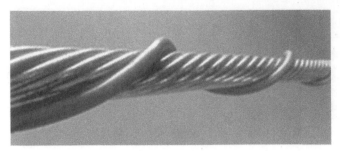

Figure 9-9   Fiber wrap.

process. Fiber systems actually operate as electronic to optical and optical to electronic modems.

Fiber cables can be wrapped around existing static wires very easily. Many existing transmission lines incorporate *fiber wrap* technology, mainly on the shield wires as shown in Figure 9-9.

A typical fiber cable terminations cabinet in a substation control building is shown in Figure 9-10. Each strand has a fiber connector. Thick jackets are used around each fiber strand for protection.

Microwave Radio

Microwave radio (MW) communications systems like those shown in Figure 9-11 use special *parabolic shaped reflector antennas* (called *dishes*) to reflect radio energy coming out of the *feedhorn* into a beam pointing toward the MW receiver antenna. These super high frequency (SHF) *line of sight radio waves* travel through air at near the speed of light. The receiving antenna at the opposite end of the radio path reflects the energy into another feedhorn where the waveguide transports the radio energy to the communications receiver. The nature of microwave energy enables the

Figure 9-10   Fiber termination.

Figure 9-11   Microwave communications.

use of narrow rectangular waveguides to transport the SHF radio energy between the radio equipment and dish antennas. These point-to-point microwave communications systems can span distances of up to about 100 km without repeaters and can communicate analog or digital data, voice, and video signals.

The drawing in Figure 9-12 shows how the SHF radio signals bounce off the reflected dish antennas and travel down waveguides to the radio equipment. The microwave radio at both ends has both receivers and transmitters. The systems operate on two different frequencies so that two-way simultaneous communication is possible.

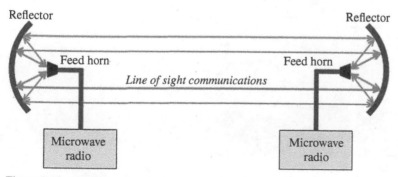

Figure 9-12   Microwave systems.

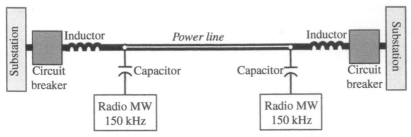

Figure 9-13   Power line carrier.

Power Line Carrier

Power line carrier systems (PLC) operate by superimposing a high-frequency radio signal (150–300 kHz) onto an existing low-frequency power line (60 Hz). These systems are point-to-point (i.e., substation-to-substation). They offer slow data rates compared to fiber or microwave systems. PLC systems like that shown in Figure 9-13 have been in operation for many decades and several systems are still in service today. The communication path is typically the center phase of the three-phase transmission line.

Referring to Figure 9-14, the theory of operation takes into account the fact that high-frequency radio signals pass easily through capacitors yet are blocked or

Figure 9-14   PLC system.

severely attenuated by inductors or coils. Whereas, low-frequency signals are just the opposite, they pass through inductors easily yet are blocked by capacitors. The drawing below shows how the equipment is located on a power line between substations.

The inductors are sometimes called *line traps* or *wave traps* and the capacitors are called *coupling capacitors*. *Notice* that the radio communications occur between the coils and the circuit breakers. Therefore, a line fault that trips the circuit breakers still enables communications (unless the line is cut).

There are a few drawbacks to this older PLC technology, such as transformers severely attenuate PLC signals, snow and rain weather conditions can cause high noise levels and high noise causes data errors. Therefore, PLC has bandwidth limitations when used for Internet or other high-speed communications services.

Radio Communications

Multiple address systems (MAS) are point-to-multipoint (P-MP) networks that are usually configured in a "star" architecture. They communicate to multiple fixed remote radio stations, one at a time. In each area, one "master" station communicates with multiple RTUs, usually in the 450, 900, or 1400 MHz radio frequency spectrum (referred to as the "UHF" or ultra-high-frequency band). MAS systems can be designed for either licensed or unlicensed use, which is usually decided based on the criticality of the equipment to be controlled or monitored.

Point-to-point (P-P) and *point-to-multipoint* (P-MP) radio communications systems are used by electric utilities for many reasons. P-MP systems are commonly used to provide SCADA data communications services between system control centers and SCADA remote terminal units, usually when fiber optics or microwave radio is too costly. P-MP radio systems are also used as base station systems to communicate with field crews. Note, portable P-P radio systems are used for voice communications but are quickly being replaced with cellular phone technology systems.

The most common form of communications used for distribution automation, SCADA, and distribution feeder communications is licensed or unlicensed UHF wireless communications.

Copper Communications

Electric utilities might use *twisted pair copper* communications systems between substations for protective relaying applications.

There are basically two ownership scenarios involved in copper communications systems; the utility can own the copper cables or the copper circuits are leased from a third party, such as the local telephone company. Leased circuits are used when there are low-priority applications, such as voice, remote metering, and interruptible load control. Whereas, leased circuits are not normally used for high reliability data circuits, such as SCADA or system protection. Electric utilities prefer using privately owned in-house copper cable circuits for critical data communications since they have full control of emergency maintenance and reliability issues.

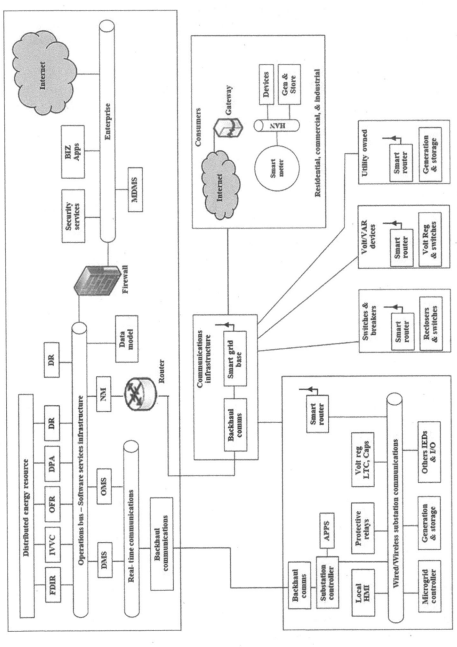

Figure 9-15   Distribution management. Courtesy of John McDonald and Mini Thomas (redrawn).

Satellite Communications

Satellite communications are used in electric power systems for applications that can tolerate the inherent 2 second delay times. For example, meter reading and remote information monitoring work well with satellite communications. High-speed protective relaying applications do not bode well with satellite communications because the inherent time delay is intolerable. Also, satellite voice communications have pauses which can degrade quality of service.

UTILITY COMMUNICATIONS SYSTEMS

Most electric utilities use a mixture of the above described communications systems. The communications channels used by utilities can be provided by the utility itself or be leased through third-party communications service providers. Typically, utilities own their communications systems in order to have better control over repair response and preventative maintenance since this function is very critical, especially when used for protective relays, system control centers, and special/secure corporate voice and data applications.

Communication equipment used in the electric power grid system must be able to provide service to several applications, such as SCADA, protective relaying, corporate communications, distribution automation, customer information services, and more. The components used in modern power utility communications systems include digital interfaces for Intelligent Electronic Devices (IEDs), Programmable Logic Controllers (PLCs), Wide Area Monitoring Systems (WAMS), Smart Grid (SG) equipment, plus protective relaying, sequence of events recorders, GPS time synchronization, PMUs, state monitoring servers, and analog-digital converters (merging units). The demand and dependability utilities have on modern telecommunications infrastructure and interfaces are growing daily.

Figure 9-15 shows the sophistication of a modern power utility communications system. The integration and connectivity of all the new smart grid operations and management technologies into the existing communications essentials of SCADA, EMS, and protective relays shows the growing need of elaborate, reliable, dependable, and secure utility telecommunications systems.

CHAPTER 10

PERSONAL PROTECTION (SAFETY)

CHAPTER OBJECTIVES

After completing this chapter, the reader will be able to:

- ☑ *Discuss "Personal Protection Equipment" used for safety in electric power systems*
- ☑ *Explain human vulnerability to electricity*
- ☑ *Explain how one can be safe by "Isolation" or "Equipotential"*
- ☑ *Discuss "Ground Potential Rise" and associated "Touch" and "Step" potentials*
- ☑ *Discuss how "Energized" or "De-energized and Grounded" lines provide safe working environments for field workers*
- ☑ *Discuss the "Safety Hazards" around the home*

ELECTRICAL SAFETY

The main issues regarding *electrical safety* are the invisible nature of hazardous situations and the element of surprise. One has to anticipate, visualize, and plan ahead for the unexpected and follow all the proper safety rules before an accident to gain confidence in working around electricity. Those who have experience in electrical safety must still respect and plan for the unexpected. There are several methodologies and personal protective equipment (PPE) available that make working conditions around electrical equipment safe. The common methodologies and safety equipment are explained in this chapter. The theories behind those methodologies are also discussed. Having a good fundamental understanding of electrical safety principles is very important and is effective in recognizing and avoiding possible electrical hazards.

There are two aspects of electrical safety that are discussed in this chapter; *electric shock* or current flow through the body and *arc-flash* or being burned by the heat created by an electrical arc when equipment failure occurs. Protection against electrical shock is discussed first.

Electric Power System Basics for the Nonelectrical Professional, Second Edition. Steven W. Blume.
© 2017 by The Institute of Electrical and Electronics Engineers, Inc. Published 2017 by John Wiley & Sons, Inc.

PERSONAL PROTECTION

Personal protection refers to the use of proper clothing, insulating rubber goods or other safety tools that provide *electrical isolation* from electrical shock. Another form of personal protection is the application of *equipotential* principles, where everything one comes in contact with is at the same potential. Electrical current cannot flow if equipotential exists. Either way, using insulating personal protection equipment (PPE) or working in a *zone of equipotential* are known methods for reliable electrical safety.

Human Vulnerability to Electrical Current

Before discussing personal protection in greater detail, it is helpful to understand human vulnerability to electrical current. The level of current flowing through the body determines the seriousness of the situation. Note, the focus is on current flow through the body opposed to voltage. Yes, a person can touch a voltage, create a path for current to flow, and experience a shock, but it is the current flowing through the body that causes issues.

Testing back in the early 1950s showed that a range of about 1–2 mA (0.001–0.002 A) of current flow through the human body is considered the threshold of sensitivity. As little as 16 mA (0.016 A) can cause the loss of muscle control (*lock-on*). As little as 23 mA (0.023 A) can cause difficulty breathing, and 50 mA can cause severe burning. These current levels are rather small when compared to normal household electrical load. For example, a 60 W light bulb draws 500 mA of current at full brightness with rated voltage of 120 V.

The residential *ground fault circuit interrupter* (GFCI) like those used in bathrooms (discussed earlier) open the circuit if the differential current reaches approximately 5.0 mA (0.005 A). The GFCI opens the circuit breaker before dangerous current levels are allowed to flow through the human body. The conclusion is humans are very vulnerable to relatively small electrical currents.

Principles of "Isolation" Safety

A person can be safe from electrical hazards through the use of proper rubber isolation products, such as gloves, shoes, blankets, and mats. Proper rubber goods allow a person to be isolated from *touch and step potentials* that would otherwise be dangerous. (Note, touch and step potentials are discussed in more detail later in this chapter.) Electric utilities test their rubber goods frequently to insure that safe working conditions are provided.

Rubber gloves are routinely used when working on de-energized high-voltage equipment just in case it becomes accidently energized. Rubber gloves are also used for hot-line maintenance at distribution voltage levels only. Figure 10-1 shows the cotton inner liners, insulated rubber glove, and leather protector glove used in typical live line maintenance on distribution systems or to protect against accidental energization.

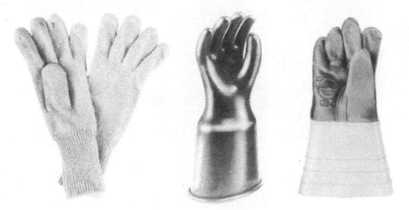

Figure 10-1   Rubber gloves. Courtesy of Alliant Energy.

Figure 10-2 shows high-voltage *insulated boots*. Figure 10-3 shows typical high-voltage *insulated blankets and mats*. Every electric utility has extensive and very detailed safety procedures regarding the proper use of rubber goods and other safety-related tools and equipment. Adherence to these strict safety rules and equipment testing procedures insures that workers are safe. Further, electric utilities spend generous time training workers to work safety, especially when it comes to live line activities.

Principles of "Equipotential" Safety

Substations are built with a large quantity of bare copper conductors and ground rods connected together and buried about 18–26 inches below the surface. Metal fences,

Figure 10-2   Insulated boots.

Figure 10-3   Rubber blankets and mats. Courtesy of Alliant Energy.

major equipment tanks, structural steel, and all other metal objects requiring an electrical ground reference are all connected to the buried copper conductors. This elaborate interconnected system of conductive metals form what is referred to as the station *ground grid*.

This elaborate ground grid provides a safe working environment that is sometimes referred to as *equipotential grounding*. Usually a copper conductor is buried outside the fence perimeter (approximately 3 feet from the fence) to extend the ground grid for additional safety. Usually 2–4 inches of clean gravel is placed on top of the soil in the substation to serve as additional isolation from current flow and voltage profiles that could exist in the soils during fault conditions. Figure 10-4 shows the ground grid concept.

There are two main reasons for having an effective grounding system; first, to provide a highly effective ground path through earth soil for fault current to flow back to the source in order to trip circuit breakers (i.e., system protection). Second, effective grounds provide a zone of equipotential for safe working environments (i.e., personnel protection). The effective ground grid causes high fault currents to trip

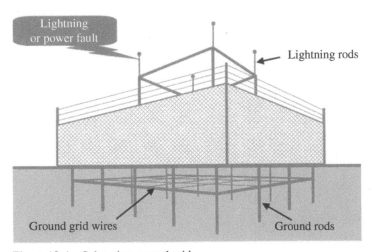

Figure 10-4   Substation ground grid.

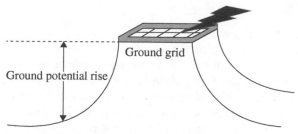

Figure 10-5   Substation ground potential rise.

circuit breakers faster. The zone of equipotential minimizes the risk of someone experiencing a current flow during a lightning strike or power fault. Theoretically, everything a person touches in a *zone of equipotential* is at the same voltage and therefore no current flows through the person. As an example, suppose you were in an airplane flying above the earth at 30,000 feet. Everything inside the airplane seems normal. The same is true in a properly designed substation when a 30,000 V *ground potential rise* occurs.

Ground Potential Rise

When a fault occurs on a power system, a *ground potential rise* (GPR) condition occurs where high electrical currents flow in the earth soil creating a voltage profile on the earth's surface. This voltage profile decays exponentially outward from the fault location as shown in Figures 10-5 and 10-6. This GPR condition can cause dangerous *touch and step potentials*.

The following drawings show the GPR and touch and step potentials.

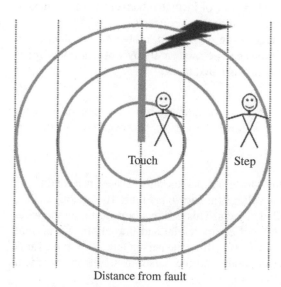

Distance from fault

Figure 10-6   Touch and step around structures.

Touch and Step Potentials

During a lightning strike or power fault event in a substation, the entire substation rises to a high potential and anyone standing on the ground grid during that event should experience no touch or step potential because of the equipotential grounding. *Touch potential* is the difference between the voltage magnitude of a person (or animal) touching an object and the magnitude of voltage at the person's feet. Touch potential can also be the difference in voltage between two potentials (i.e., hand to hand). *Step potential* is the difference in voltage between a person's (or animal's) feet. Shoes, gloves, and other articles of clothing help insulate a person from touch and step potentials. Approved, tested, and properly used rubber safety products (PPE) provide isolation from potentially hazardous touch and step potentials.

Working Transmission Safely

Construction and maintenance crews work on power lines under-energized and de-energized conditions. Either way, special safety precautions are mandatory. All safety precautions fall back to the basic principles of either being fully isolated from electric shock or be in a zone of equipotential. One has to plan on the possibility of a de-energized line becoming accidently energized without notice. Following are examples of different ways to work on power lines safely.

Energized Equipment

There are multiple ways to work on energized power lines safely; insulated bucket trucks, the use of fiber glass non-conductive hot sticks and bare hand live line maintenance are the more common means.

Insulated Bucket Trucks   Working out of insulated bucket trucks is a means of working on lines that are either energized or de-energized.

Depending on the system voltage being worked on, rubber gloves, fiber glass hot sticks, or live line bare hand methods can be used safely by working out of *insulated trucks*. Figure 10-7 shows using an insulated truck

Hot Stick Live Line Maintenance   Work can be performed when the lines are energized using hot sticks. Figure 10-8 shows workers using fiberglass *hot sticks* to perform maintenance.

"Bare Hand Live Line" Maintenance   A person can be placed in a conductive suit and touch energized transmission voltages as shown in Figure 10-9 as long as they do not come in contact with grounded objects. This is like a bird sitting on the wire. The *conductive suit* establishes a zone of equipotential and thus eliminates current flow inside the suit or human body. Since everything the person touches is at the same potential, no current flows through the body and the person is safe from electrical shock. (Picture of author touching live 345 kV)

Figure 10-7   Insulated buckets.

De-energized Equipment and Grounded

During de-energized conditions, workers apply *ground jumpers* to avoid danger-ous potentials, should the line become accidentally energized. Grounding equipment serves two purposes:

1. Grounding establishes a safe zone of *equipotential* similar to substations. It provides a safe environment against "touch potentials."

2. Grounding helps trip circuit breakers faster, should the line become accidentally energized.

Figure 10-10 shows several jumpers on a rack waiting to be used on a power line or substation.

Working Distribution Safety

Similar to transmission line work, distribution line crews work under-energized or de-energized conditions also. Special safety procedures are mandatory in either situation. Distribution line crews work energized lines (normally under 34 kV) using *rubber isolation equipment* (PPE) (i.e., rubber gloves and blankets) for voltages less than

Figure 10-8   Live maintenance transmission lines.

Figure 10-9   Bare hand live line maintenance.

Figure 10-10   Ground jumpers.

usually 34 kV. Figure 10-11 shows live line maintenance activities on distribution systems. Working lines de-energized requires "ground jumpers" as discussed earlier.

Switching

Switching is the term used to change the configuration of the electric system or to provide isolation for safe working activities on equipment, such as maintenance. Switching is required to open or close disconnect switches, circuit breakers, etc. for

Figure 10-11   Live maintenance distribution. Courtesy of Alliant Energy.

Figure 10-12   Live maintenance substations.

planned maintenance, emergency restoration, load transfer, and equipment isolation. Figure 10-12 shows a switching event in an energized substation. Switching requires careful control of all personnel and equipment involved. This usually requires radio, phone, or visual communication at all times for safety assurance. Detailed radio and equipment *tagging procedures* are also required to help prevent others from interfering with work activities. Switching can be very time consuming due to the repetitive nature of the communication of the *switching orders*.

ARC-FLASH

Electrical *arc-flash* hazards are serious risks to worker safety. On the average, every day in the United States, five to ten people are sent to special burn units due to arc-flash burns. The National Fire Prevention Association (NFPA) published NFPA 70E®, the standard for Electrical Safety in the Workplace®, in order to document electrical safety requirements regarding arc-flash safety. 70E® defines specific rules for determining the category of electrical hazards and the PPE required for personnel working in defined and marked hazard zones or boundaries. OSHA enforces the NFPA arc-flash requirements under its "general rule" that a safe workplace must be maintained. These regulations are forcing employers to review and modify their electrical systems and work procedures to reduce the arc-flash hazard and to improve worker awareness and safety.

Industry regulations and standards now require the electrical equipment owner to do the following:

- Assess whether there are arc-flash hazards
- Calculate the energy released by the arc, if or when present
- Determine the flash protection boundaries

- Provide appropriate PPE for personnel working within the flash protection boundary
- Provide a safety program and training with clear responsibilities
- Suitable tools, in addition to PPE for a safe workplace
- Equipment labels indicating the minimum protective distance, the incident energy level, and required PPE for that location

Employees too have an obligation to arc-flash safety. They must follow the requirements of arc-flash labeling by wearing the proper PPE and use proper safety tools provided by their employer. Further, they must not work on or near electrical circuits or equipment unless they are "qualified" workers.

About the Arc

An arc-flash is the light and heat produced from an electric arc when supplied with sufficient electrical energy that can cause substantial damage, harm, fire, and/or serious injury. Electrical arcs, when controlled, produce a very bright light and when controlled, can be used in arc lamps (having electrodes), for welding, plasma cutting and other industrial applications.

When an uncontrolled arc forms at high voltages, arc-flashes can produce deafening noises, supersonic concussive-forces, super-heated shrapnel, temperatures far greater than the sun's surface, and intense high-energy radiation capable of vaporizing nearby materials. Arc-flash temperatures can reach or exceed 35,000°F (19,400°C) at the arc terminals. The result of the violent event can cause destruction of equipment, fire, and injury not only to an electrical worker but also to bystanders. Figure 10-13 shows what happens after an arc-flash event.

Figure 10-13   Arc-flash. Reproduced with permission of from LSelectric.

Hazard Categories

NFPA 70E® includes *hazard categories* that take into account human vulnerability factors and the capability of available PPE clothing used to protect humans that are exposed to arc-flash incidents. Calories/cm² is the reference used in arc-flash criteria. One calorie/cm² can be equal to holding your finger over the tip of a flame of a cigarette lighter for one second. Specifically, one calorie is the amount of heat needed to raise the temperature of one gram of water by 1°C. Further, thermal energy is rated in calories/cm². A second-degree burn requires approximately 1.2 cal/cm² for more than one second.

Below are the categories used in arc-flash rules and regulations. These categories correspond to required PPE.

- Category 0: Up to 1.2 cal/cm²
- Category 1: 1.2–4 cal/cm²
- Category 2: 4.1–8 cal/cm²
- Category 3: 8.1–25 cal/cm²
- Category 4: 25.1–40 cal/cm²
- Over 40 cal/cm²: Unacceptable risk

Protective Clothing and Equipment

Personal protective equipment (PPE) is the common term used for clothing and equipment to protect electrical workers performing activities on or near energized high-voltage equipment. In the case of exposure to arc-flash hazards and depending on the hazard risk category (defined by NFPA), PPE is primarily made up of flame resistant (FR) clothing. FR clothing is the most common and industry accepted PPE to protect the body from burns due to flame. It is not, however, designed to isolate the worker from electrical contact. The beneficial characteristic of FR clothing is that it will not continue to burn on its own when a flame source is removed. This protection is achieved by treating the fiber cloth with flame retardant "modacrylic" blended cottons.

Table 10-1 shows the required FR PPE required for the various hazard/risk categories.

TABLE 10-1   Required Flame Retardant Clothing

Hazard/Risk Category	Eye Protection, Ear Canal Inserts, Long Sleeve Shirt and Pants	Arc Rated Clothing	Face and Head Protection	Flash Suit Hood
0	√			
1	√	√	√	
2	√	√	√	
3	√	√	√	√
4	√	√	√	√

Figure 10-14   Arc-flash clothing. Reproduced with permission of from LSelectric.

In most applications, clothing and PPE must be either FR rated or arc-flash rated. The electrical worker should never wear materials such as nylon and polyester that can melt and stick to skin. Note non-FR-rated undergarments may catch fire even when arc rated clothing is worn overtop and survives an arc-flash.

Although some of this clothing and equipment appears to be bulky, restraining, and cumbersome, manufactures try hard to add flexibility, light weightiness, and durability in their FR clothing while meeting the strict arc-flash requirements. Figure 10-14 shows the clothing required for Category 4 energy.

The arc rating of the FR material is the maximum incident energy resistance demonstrated by a material prior to break open (a hole in the material) or to pass through and cause a second- or third-degree burn with 50% probability.

Arc rating is normally expressed in cal/cm^2 (or small calories of heat energy per square centimeter). The tests for determining arc rating are defined in ASTM F1506 Standard Performance Specification for Flame Resistant Textile Materials for Wearing Apparel for Use by Electrical Workers Exposed to Momentary Electric Arc and Related Thermal Hazards.

Approach Boundaries

Based on the hazard categories stated earlier and the PPE flame retardant capabilities, NFPA 70E® also stipulates four arc-flash approach boundaries that must be known and observed. Figure 10-15 shows these approach boundaries.

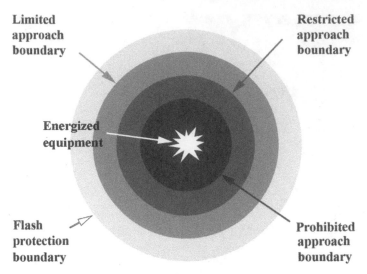

Figure 10-15   Arc-flash boundaries.

Flash protection boundary (outer boundary) is where a worker is exposed to a curable second-degree burn if outside this boundary.

Limited approach boundary is the closest area an unqualified person can safely stand. Still requires PPE. A person must be qualified to go any closer.

Restricted approach boundary is the closest a qualified person can stand, provided they have an approved plan for the work to be performed and it must be absolutely necessary to work in this area.

Prohibited approach boundary (inner boundary) is considered the same as making contact with the energized part. Requires qualified person with specific training to work on energized conductors, appropriate PPE, and documented plan with justification.

As you can see, electrical workers are required to work safely using proper PPE for both electrical contact and arc-flash hazards. All industries have their hazards; the electrical industry has these two concerns. Proper training, safety equipment, and having the proper understanding of the hazards before a live contact or arc-flash event occurs provide electrical workers with the necessary situational awareness, tools, and assurance that a safe working environment is provided.

ELECTRICAL SAFETY AROUND THE HOME

Home safety also involves the awareness of touch and step potentials, arc-flash too. Whether one is exposed to a dangerous touch or step potential in a substation or at home, the same circumstances exist and the same precautions are necessary. As soon as the insulation around energized wires is compromised, dangerous step and touch potentials can exist, plus a bright arc-flash event can occur if and when frayed insulated conductors touch.

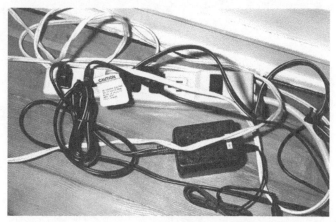

Figure 10-16   Safety at home. Reproduced with permission of Photovault.

Figure 10-17   GFCI receptacle. Reproduced with permission of Photovault.

For example, worn extension cords can have exposed conductors that can cause 120 Vac touch potential hazards. All worn cords must be replaced. To compound the problem, water, moisture, metal objects, and faulty equipment can increase the possibility of injury from accidental contact. Everyone is vulnerable to electrical current, therefore always be vigilant about electrical safety at home!

Figures 10-16 and 10-17 show how electrical safety starts at home. Reproduced with permission of Photovault.

Always be vigilant about electrical safety at home!

APPENDIX *A*

THE DERIVATION OF ROOT MEAN SQUARED

Since the average value of voltage or current in an ac sine wave is zero, the average value of each half of the sine wave is calculated and added together to determine its total effective average value. This total effective average value of an ac sine wave has the same heating effect to that of dc circuits. The process of finding the total *effective* average value of a sine wave is a method called *root mean squared*, or *rms*.

The rms value of voltage and current is shown in Figure A-1 below:

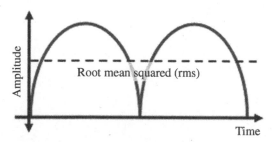

Figure A-1 Root mean squared.

Residential Voltage

The 120 Vac value of residential voltage is stated in rms. Further, high voltage distribution and transmission lines are also stated in rms. Note, multiplying the rms value by the square root of 2 produces what is known as the *peak* value. In the case of residential voltage, the peak value is 165 Vac. Multiplying the peak value by 2 results in the term called the *peak-to-peak* value, which is the total measurement of the

Electric Power System Basics for the Nonelectrical Professional, Second Edition. Steven W. Blume.
© 2017 by The Institute of Electrical and Electronics Engineers, Inc. Published 2017 by John Wiley & Sons, Inc.

magnitude of the sine wave as can be seen on an *oscilloscope*. (An oscilloscope is a visual voltage measuring device.)

Residential voltage

$V_{rms} = 120$ Vac

$V_{peak} = 165$ Vac

$V_{peak-peak} = 330$ Vac

GRAPHICAL POWER FACTOR ANALYSIS

SOMETIMES IT IS EASIER to understand the relationship between real and reactive power graphically. Basically, resistors dissipate energy in the form of heat while performing work functions. The power associated with resistive loads is expressed as *Watts*. The reactive power associated with capacitive and inductive loads is expressed as *VARs*. Reactive power, VARs is *Watt-less* power and does not contribute to real work functions. Reactive power (VARs) is required in motors, transformers, and other ac functioning coils to produce magnetic fields. These magnetic fields are needed to make motor shafts spin; however, the real work done by motors is the load placed on its spinning shaft. The total power supplied to an inductive load such as a motor is the Watts plus the VARs.

One interesting fact that exists in ac power systems is that inductive VARs are opposite of capacitive VARs and can cancel each other out if they are the same value.

The graphical means of showing the relationship between the real and reactive power associated with resistors, inductors, and capacitors is shown in Figure B-1.

Note how the inductive and capacitive VARs oppose each other and can cancel, yet resistive Watts remain independent.

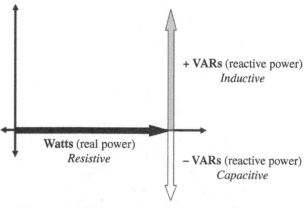

+ VARs (reactive power)
Inductive

Watts (real power)
Resistive

– VARs (reactive power)
Capacitive

Figure B-1   Electrical power relationships.

Electric Power System Basics for the Nonelectrical Professional, Second Edition. Steven W. Blume.
© 2017 by The Institute of Electrical and Electronics Engineers, Inc. Published 2017 by John Wiley & Sons, Inc.

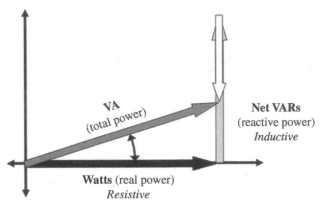

Figure B-2   Power triangle.

Figure B-2 shows the power triangle with the capacitive VARs cancelling most of the inductive VARs. The result is net VARs. In this example, the net VARs are still positive (i.e., the circuit remains inductive). Not all of the inductive VARs were cancelled by capacitive VARs.

The hypotenuse *VA* represents the *total* power; sometimes referred to as the *apparent* power. Total power or apparent power is the peak voltage times the peak current.

For there to be a power factor angle, the current must lead or lag voltage. This figure shows net inductive load. In this case, the real part of the VA (Watts) is the peak voltage of the sine wave times the current at the time of peak voltage.

The *power factor angle* shown in the graph is the same lagging phase angle shown earlier in Figure 6-2.

RECOMMENDED READING

"2014 Renewable Energy Data Book," U.S. Department of Energy, Energy Efficiency & Renewable Energy, www.energy.gov

"2016 Modern Solutions Catalog," Schweitzer Engineering Laboratories, Inc., www.selinc.com

"Arc Flash Hazards, Low Voltage Circuit Breakers," ABB Inc., www.abb.com

ASTM F1506, "Standard Performance Specification for Flame Resistant and Arc Rated Textile Materials for Wearing Apparel for Use by Electrical Workers Exposed to Momentary Electric Arc and Related Thermal Hazards,"2015.

"EPRI Power System Dynamics Tutorial," Electric Power Research Institute, 2009.

IEEE Standard 1159, "IEEE Recommended Practice for Monitoring Electric Power Quality," 2009.

IEEE Standard C37.2, "Standard for Electrical Power System Device Function Numbers," 2008.

Mini S. Thomas and John D. McDonald, *Power System SCADA and Smart Grids*, CRC Press, Taylor & Francis Group, 2015.

National Renewable Energy Laboratory, www.nrel.gov

NFPA 70E, "Standard for Electrical Safety in the Workplace®," National Fire Prevention Association, 2003.

"Reliability Standards," North American Electric Reliability Corporation, www.nerc.com

Electric Power System Basics for the Nonelectrical Professional, Second Edition. Steven W. Blume.
© 2017 by The Institute of Electrical and Electronics Engineers, Inc. Published 2017 by John Wiley & Sons, Inc.

INDEX

Electric Power System Basics for the Nonelectrical Professional, Second Edition. Steven W. Blume.
© 2017 by The Institute of Electrical and Electronics Engineers, Inc. Published 2017 by John Wiley & Sons, Inc.

IEEE Press Series on Power Engineering

Series Editor: **M. E. El-Hawary,** Dalhousie University, Halifax, Nova Scotia, Canada

The mission of IEEE Press Series on Power Engineering is to publish leading-edge books that cover the broad spectrum of current and forward-looking technologies in this fast-moving area. The series attracts highly acclaimed authors from industry/academia to provide accessible coverage of current and emerging topics in power engineering and allied fields. Our target audience includes the power engineering professional who is interested in enhancing their knowledge and perspective in their areas of interest.